ENVIRONMENTAL POLITICS

Interest Groups, the Media, and the Making of Policy

ENVIRONMENTAL POLITICS

Interest Groups, the Media, and the Making of Policy

NORMAN MILLER

LEWIS PUBLISHERS

A CRC Press Company
Boca Raton   London   New York   Washington, D.C.

Library of Congress Cataloging-in-Publication Data

Miller, Norman, 1939-
  Environmental politics : interest groups, the media, and the making of policy
  / Norman Miller
      p. cm.
  Includes bibliographical references and index.
  ISBN 1-56670-522-5 (alk. paper)
  1. Environmental Policy. 2. Environmental protection—Government policy.
  I. Title.

GE170.M54 2001
363.7′056—dc21   2001038065
                                                                        CIP

This book contains information obtained from authentic and highly regarded sources. Reprinted material is quoted with permission, and sources are indicated. A wide variety of references are listed. Reasonable efforts have been made to publish reliable data and information, but the author and the publisher cannot assume responsibility for the validity of all materials or for the consequences of their use.

Visit the CRC Press Web site at www.crcpress.com

No claim to original U.S. Government works
International Standard Book Number 1-56670-552-5 (alk. paper)
Library of Congress Card Number 2001038065
Printed in the United States of America 1 2 3 4 5 6 7 8 9 0
Printed on acid-free paper

Dedication

To Andrea, who was part of this book from the beginning

I have never regarded politics as the arena of morals.
It is the arena of interests.   **Aneurin Bevan, British Labour politician**

CONTENTS

Preface xi
Acknowledgments xv
The Author xvii

1. The Changing Landscape of Environmental Politics 1

2. Legislation: Leveling the Playing Field and Leveraging
   the Process..13

3. Environmental Regulation and the Reinvention of
   the EPA 31

4. The Media Business and Environmental Politics......................53

5. Uncertain Science—Uncertain Politics 65

6. State and Local Governments: The "Other"
   Interest Groups..81

7. The Shifting Tides of Environmental Advocacy.......................91

8. The Greening of Business: Politics for Profits 111

9. Global Pressures and Domestic Environmental Politics.........129

10. On-Line Activism and the New Environmental Politics 137

Bibliography 153

Index 163

PREFACE

Every book is, at its roots, a personal narrative, an attempt to rationalize the experience of its author. This book is no exception. It was conceived long before I sat down at my computer to try to draw some overarching conclusions about what in my experience as a foot soldier in the environmental wars might be interesting and instructive to others. But it did not come together until several years after I had left the field of battle and had the opportunity to undertake some mental reconnaissance flights over the theater of war, when I came to understand the relation between the battles and the war, to determine which strategies succeeded and which failed, to sort out the victors and the vanquished, and to try to anticipate to the extent possible what battles lie ahead and what weapons may be needed to fight them effectively.

More than 25 years ago, as a result of a fortuitous random draw, I was assigned to the Environment Section of the New Jersey legislature's research staff. I say fortuitous because I did not know then, nor could I have known, that environmentalism was destined to become the most significant sociopolitical movement of the last half of the twentieth century, that it would generate an unprecedented body of law, and that it would be responsible for the continuing expenditure of vast sums of public and private revenues and change our culture and our view of our world and ourselves. Nor did I anticipate that New Jersey, by virtue of its dense population, extensive coastline, and vast farmlands and woodlands juxtaposed to major petrochemical, pharmaceutical, and oil refining facilities would be among the first states to confront the complex problems and issues that came to characterize this new movement.

But I was innocent as well of just how public policy was formulated. I brought to my new assignment the only kind of understanding that people who have not been intimately involved in policymaking can have, a civics textbook notion that laws are the product of the informed deliberation of elected, albeit partisan, officials and technically proficient administrative personnel and that these laws are implemented effectively by governments at all levels for the "public good."

But my experiences were an assault on these preconceptions. I was shocked when, at a committee meeting I staffed on the Clean Air Act, the first three rows of seats were commandeered by members of classic and antique car clubs, there to oppose the stricter regulations justifiably imposed on older, more polluting automobiles. I was distressed when television reporters were reluctant to cover what we regarded as a monumental public interest story—the contamination of the almost 50% of the state's potable water supply by leaking underground gasoline storage tanks—because the story could not have a dramatic visual vehicle. Underground tanks are, after all, impossible to film.

Over the years, I looked on as education officials participated actively in the consideration of legislation reducing real estate taxes adjacent to riparian land because these monies were dedicated to securing school bond funds; as licensed well-drillers welcomed regulations that would cause them only minor inconvenience and cost, but that would drive out competition from incompetent practitioners and moonlighters; as farmers failed to support farmland preservation programs because they wanted to preserve their options as speculators; and as animal rights activists split with their environmental friends over the testing of health hazards of suspect chemicals because such tests would be conducted on animals. And, as staff, we struggled to bring the expertise of the state's scientific community to the committee room, a community that needed to preserve its options as advisers to industry, advocacy groups, and governments, not to the general public.

I would like to return to the military imagery with which I began, for it is consistent with three basic truths that I discovered during the course of my work in environmental politics. The first is that policies are more often the product of combat than deliberation. The second is that the combatants are multiple and varied, do not always wear uniforms to identify their missions, and change sides when it suits their purpose to do so. The third truth is that winning strategies in the present climate depend as much on activities off the designated battlefield as on it. This book is an effort to explain the environmental policy consequences of these truths.

The first two of these truths are stressed throughout the text because they constitute what I see as the driving principles in environmental policymaking. But I would like to briefly reiterate the third, since its manifestation is more subtle. A central thesis of this book is that the past several decades have witnessed an evolution in environmentalism from a high-minded conservation movement to a populist, grassroots social force, with a corresponding shift in the locus of advocacy from the corridors of government to the broad social environment we all inhabit daily. Thus, the regular columns of Exxon-Mobil that appear several times weekly in the *New York Times*, the stream of letters to the editor in general-circulation publications in response to environmental stories, the websites tacked on to television spots sponsored by advocacy groups on the relevant topics of the day, and even the packaging designed to attract ecology-minded consumers all constitute part of what has become the politics of the environment. Public opinion is now more influential than ever, and the appeals to it are all around us. I hope this book encourages its readers to become more sensitive to these subtle, often subliminal, lobbying efforts. Students may well want to make a challenge out of seeking them out and placing them in their appropriate context. Environmentalism has insinuated itself into the fabric of our culture and lifestyle, and thus courses in sociology and psychology as well as environmental studies may use this book profitably as a text.

In a larger context, though this book examines the policymaking process through the lens of the environment, it should be understood that that process is not fundamentally different from the policy process that shapes other public issues such as health, labor, or civil liberties. Thus the book could well serve as a text in political science curricula, with environment treated simply as a case in point.

Finally, although the book is targeted to those with an academic interest in the subject, I feel strongly that it would be useful to the work of people in positions in both the public and private sectors whose responsibilities involve them in issues that are addressed by proposed legislation or regulation that affects them or their employers—to bankers concerned about the transfer of property previously implicated in chemical operations; to farmers anxious about pending government regulation of pesticides and herbicides; to food marketers potentially affected by how genetically modified food products will be handled by governments here and abroad; by employers trying to ensure safe and healthy working environments for their employees within economically manageable frameworks; and for the many professionals and trade unionists who get caught up in large social issues that get entangled with theirs, even though indirectly.

In sum, I believe that this book provides a broad perspective for anyone—government official, business manager, formal advocate, or concerned public-spirited citizen—who wants to move public policy in a direction that he or she views as personally or socially desirable. Each of them, knowingly or not, has become a constituent of one or another of the interest groups that matter now more than ever. Their power has never been greater.

ACKNOWLEDGMENTS

As I was writing the final chapters, I couldn't help thinking of Huck Finn's lament at the end of *his* personal narrative: ". . . if I'd a knowed what a trouble it was to make a book I wouldn't a tackled it. . . ." Huck was right; it *is* a trouble to make a book. But, unlike Huck, I had some first-class assistance, without which I would have been . . . well . . . up a creek.

First, I want to acknowledge the assistance of Michael Catania, who reviewed each chapter with his keen eye for accuracy and nuance. While his comments, suggestions, and corrections are most explicitly represented in the final draft, there is much of him in its core content; as a colleague of mine for many years, he taught me more about environmental policy than he is aware of.

I want to thank, as well, a former student of mine at Rollins College who graciously volunteered to help me through the production of the book, no mean task considering its author's technophobia. Barbara Howell took on responsibilities for the "art" (what publishing houses euphemistically call *technology*), for introducing me to "Zip diskography," for seamlessly transforming image from "homepage" to book page, and for literally keeping my computer functional at critical points in the writing. But much earlier, she and her fellow students over the next several years unwittingly played much more substantive roles as subjects for the test marketing of many of the ideas in this book and for helping to sharpen them by asking those annoying questions that bright students are disposed to do.

I would also like to recognize the contribution to this book of Arline Massey, late editor at CRC Press, on whose desk the proposal for it first landed. Her immediate enthusiasm for the project, and the advice and encouragement she provided to this first-time author, gave me the confidence to surmount those early obstacles. I very much regret that she could not see the final product. I want to note, as well, the continuing help of David Packer and Randi Gonzalez at CRC Press, who, even as they were constructively responsive to my questions and concerns, gave me the freedom to do it my way—as long as it was on their schedule.

In dedicating the book to my wife, Andrea, I am performing no routine act of marital obligation. During long hikes in woods, on otherwise tiring car trips, and at more dinner table conversations than we care to count, we discussed the stuff of which this book is made. She has always had the remarkable ability to balance incisive critique with unflagging support; she never let me off easy, but, then, she never let me fall off either. Needless to say, this book is much better for those conversations, and I am grateful.

The Author

Norman Miller began his environmental career as a Research Associate for the New Jersey Office of Legislative Services, where he learned about the environment from the ground up—by drafting bills, staffing standing reference committees, researching issues, and negotiating initiatives on behalf of legislative sponsors. He became Chief of the Environment Section in 1982.

After serving the legislature for a decade and a half, Miller decided, as he puts it, to "take a different seat in the stadium." He moved to the executive branch, where as Assistant Commissioner for External Affairs and then as Director of Legislative and Intergovernmental Affairs for the Department of Environmental Protection, he took up the department's mission at a time when environmentalism was first being challenged at both the national and state levels. His responsibilities involved managing the DEP's legislative portfolio; reconciling its agenda and regulatory practices with federal mandates, local government resources, and private sector demands; supervising its public outreach efforts; and directing its publications program.

Since leaving state service, Norm Miller is fulfilling a long-standing commitment to himself to share, through teaching and writing, what he learned in the more than two decades of being on the inside of environmental politics. For the past four years, he has taught a broad range of courses in the Environmental Studies program at Rollins College, including Environmental Politics, the Political Economy of Environmental Issues, the Media and Environmental Politics, Environmental Ethics, Environmental Journalism, and, lastly, Environmental Literature, which brought together his most recent career and his prior one as a college teacher of literature.

Dr. Miller is an academic member of the Society of Environmental Journalists.

Chapter 1

The Changing Landscape of Environmental Politics

> Politics: A strife of interests masquerading as a contest of principles. The conduct of public affairs for private advantage.
>
> Ambrose Bierce

In this definition from *The Devil's Dictionary*, Ambrose Bierce, one of America's most celebrated cynics, was obviously mocking the hypocrisy of our political life, but Americans have come to accept his characterization with equanimity. Today, few citizens are under any illusions that those with axes to grind or causes to pursue organize themselves and plead their special cases to lawmakers. Indeed, they regard such activities as a fundamental right of democracy. It is only when certain of these "special pleaders" gain disproportionate power and influence, especially by contributing vast sums of money to election campaigns, that the public bristles. Americans, to be sure, like a level playing field, but they generally have no problem with the game itself.

Despite their acceptance of the role of lobbyists or "lobbiers," as the representatives of special interests have been known for almost 200 years, Americans have generally regarded one or another of the major parties as the guardian of their own interests. Today, most people see politics as a battle between two camps—Republicans and Democrats—representing ideologies that are correspondingly dual and opposing—Right and Left, conservative and liberal. Such a view provides citizens who have only a passing interest in politics—which is to say, regrettably, most of our citizens—a formulation that does not often, if ever, require revision or refinement, and it allows the media to cover elections and disputes over major issues as sporting events, with all the drama and conflict inherent in such contests.

Moreover, the popular conception is that the political battlefields are almost invariably the national and state capitols, and, on the local level, town and city halls. Such a situation, many people believe, puts most of them on the sidelines and leaves participation, and even close observation, to the political professionals and those with the luxury of much spare time. These notions do not now conform to reality, if they ever did, and, for reasons we shall examine in the remainder of this book, they are decidedly not true with respect to environmental issues. Today's environmental politics are playing out all around us, every day, wherever we may reside or work, and all of us are, consciously or unconsciously, participating. It is the goal of this book to sharpen the reader's awareness of the components and forces operating in environmental politics not only to understand why and how we as a society address the problems we do, but also to empower that reader to actively take part in making those politics serve both personal and socially desirable ends.

First, it is essential to see that viewing environmental politics exclusively, or even principally, as partisan or ideological combat is not only inadequate but often misleading, even counterproductive. Consider the following:

- In 1982, Representative John Dingell, a Democrat from Michigan and Chairman of the House Committee on Energy and Commerce, joined forces with the Reagan administration in attempting, albeit unsuccessfully, to scale down many of the core provisions of the Clean Air Act. Obviously, his alliance with the major interest group in his state, the automobile manufacturers, was more compelling than loyalty to his party.
- In 1987, Alfonse D'Amato, a conservative Republican senator from New York, enthusiastically assumed a leadership role in support of the Environmental Protection Agency's newly proposed regulations toughening air quality standards. At the same time, Chicago Mayor Richard M. Daley, a lifelong organization Democrat and strong ally of the Clinton administration, opposed those standards. Senator D'Amato was more concerned with the ability of the State of New York to meet federal ambient air quality standards, which would be facilitated by imposing stringent new requirements on Midwest power plants whose emissions were wafting to the Northeast, than with the niceties of political philosophy. For his part, Mayor Daley was worried that the new regulations would disproportionately punish large metropolitan areas such as his by crippling industrial development, encouraging suburban sprawl, and promoting more auto traffic. Party politics often stops at the state's or city's boundary.
- The Chemical Industry Council of New Jersey actively supported proposed legislation granting the state intrusive and comprehensive powers not only to inspect their clients' facilities, but also to prescribe a whole range of procedures they must follow in the handling of certain

chemicals and the management of their businesses. The Toxic Catastrophe Prevention Act was introduced in the wake of the explosion at a chemical plant in Bhopal, India, the most serious industrial accident in world history. Because New Jersey has the second largest concentration of chemical plants in the U.S., opposing any measure that purported to prevent such an accident in the state would have seriously tarnished the industry's image.

■ A coalition of 52 business leaders in the West tried to block a U.S. Forest Service plan to triple commercial logging in the Sierra Nevada Mountain area north of Lake Tahoe. This departure of business from its historical alliance with the timber industry was in response to a shift in this area away from extractive industry and toward recreation and tourism. As one of the coalition businessmen put it: "We're a tourism economy now, and people come up here to see trees standing, not on the ground." Similarly, a civil war has been raging among Alaskans over whether to include the Tongass National Forest in the ban on new road-building throughout the national forest system. Some Alaskans support retaining the exception in the interest of continued timber harvesting, but others who make their living guiding tourists and sport fishermen through the rich natural areas want the increased protection that extending the ban to the Tongass would bring.

■ The Alliance for Safe and Responsible Lead Abatement, an industry group whose job is to protect drinking water from the health hazards of lead, opposed an EPA proposal to *relax* building containment requirements in favor of landfill disposal of contaminated materials. The changes would have given much of the work now conducted by the constituents in the alliance to general contractors. Further, the alliance advocated stronger, rather than weaker, regulations through grassroots appeals for public support, in the best tradition of environmental activism. Similarly, manufacturers and suppliers of environmental technologies have formed the Environmental Industry Coalition of the United States, Inc. to promote strong waste minimization and pollution prevention requirements before Congress and state and local governments. In both cases, the economic interests of the constituent members guided their political activities.

■ Two business groups have undertaken opposing campaigns in connection with the Kyoto Protocol on climate change. One —The Business Roundtable, representing 200 large companies—views the treaty as a significant economic threat, but a separate segment of that group, representing 13 blue-chip corporations, foresees economic benefits. They are probably both correct; only the outcomes of the treaty negotiations will determine how individual interests are affected.

- In a reversal of roles, a Native American tribe, the Skull Valley Band of the Goshute, is fighting the State of Utah to have its reservation designated and licensed as the site of a storage facility for half the nation's civilian nuclear waste and thus enjoy the jobs and other economic benefits that would flow from this designation. The dispute is emblematic of that in several other venues, where economically disadvantaged populations—to the chagrin of government officials and environmental justice groups—are willing to bear the risks of exposure to toxic and radioactive materials or emissions in exchange for the economic benefits that would accompany the activities that would pose them.

- In Wyoming, owners of land on which livestock and wildlife thrive and that preserves sources of clean water are battling owners of the mineral rights beneath them, where rich methane gas deposits await extraction. The water resources are desperately needed, but the revenues that the methane would generate might rescue a state with a low tax base and shrinking population from potential financial ruin.

If, as these cases demonstrate, positions on environmental issues are often unrelated to partisan or ideological considerations, the alliances that self-interest promotes are not infrequently between groups that historically have been at odds and give renewed meaning to the old saw that "politics makes strange bedfellows."

- The uneasy relationship between business and labor has nevertheless inspired at least two joint efforts. The signing of the Kyoto Protocol in 1997 brought together two old adversaries—the United Mine Workers and the Bituminous Coal Operators Association. The potential for job loss and higher energy prices that they see in the treaty has caused the two sides to temporarily put aside their differences to fight against a common enemy. More recently, another collaboration between traditional enemies, the Alliance for Sustainable Jobs and the Environment, finds its member environmental groups and labor unions overlooking their historic differences to fight off their principal threat, the World Trade Organization.

- Finally, longtime antagonists—cattlemen and environmentalists— are joining in support of land trusts as a way to protect western open spaces from development by providing a mechanism to allow cattle to graze them. Long opposed to the grazing of land, environmentalists now see "cows" as preferable to "condos." A comparable situation has inspired an unlikely alliance of environmentalists and developers, both of whom oppose a proposed mining project in

Colorado. The area around Mt. Emmons, the tentative site of the project, has seen a conversion over the last 30 years from mining and ranching to recreational activities such as hiking, camping, and skiing, on which the local economy now depends. Similarly, a coalition of environmental groups, the National Rifle Association, and a segment of the agriculture community are joining in support of a legislative initiative that would conserve tracts of farmland as open space paid for with funds in lieu of previous subsidies now being phased out.

Environmental policy positions, then, are less a function of ideology, of belief in certain core values that collectively constitute a commitment to social needs and aspirations, than of self-interest, principally, though not invariably, economic interest. Such an assertion may seem blasphemous, especially since the environment has achieved almost religious status among some segments of the population and "environmentalist" is a label that virtually everyone wears with pride. Yet the self-interest at the center of the formulation of environmental policy simply mirrors its role in our nation's political economy as a whole. It was, after all, Adam Smith, the apostle of free-market capitalism, who argued that the greater social good is best served by each person's pursuit of his or her own interests and that it is the role of our particular political system to nourish the conditions for a competition that reconciles individual and collective ends.

It is, therefore, well-grounded political theory that public policy derives from the healthy combat among competing interests, a war won presumably by those who control the political process or who have the financial resources to access it and bend it to their own purposes. Environmental policy is no exception to this strategy. What makes the politics of environmental policy-making uniquely complex and contentious, however, is that the environment as it has come to be defined at the dawn of the twenty-first century embraces an incredibly broad and diverse universe of phenomena, circumstances, and conditions, almost literally, as the word itself suggests, everything around us. Environmental concerns now reach into virtually every aspect of our lives. There is little that we as humans do that does not affect, or is not affected by, some aspect of the environment. It was not always so.

Conservationist Origins

For most of our country's history, at least until the middle of the twentieth century, the origins of what was later to become known as environmentalism resided in the conservation of natural resources—the country's oceans, rivers, lakes, and coastal zones; its parks and forests and wilderness areas; its agricultural lands; and, of course, the flora and fauna for which those resources provide habitat. The term that was applied to their protection and management was *preservation* or *conservation*, depending on whether you were an adherent of John Muir or of Gifford Pinchot, early twentieth century representatives of different schools of conservation. Their epochal struggle over the proposed utilization of the dam and reservoir at Hetch Hetchy in Yosemite National Park to meet the emergent water supply needs of San Francisco in the wake of the 1906 earthquake—the first major battle in the war of environmental politics—was over man's relationship to nature and specifically the extent to which this natural resource could or should be utilized to meet man's needs. The central issue in this debate had far-reaching philosophical and ethical implications, but it had little practical significance. Except for the residents of San Francisco, whose potable water supply needs would eventually be met by Hetch Hetchy, the dispute barely impacted the everyday lives of most Americans.

For the better part of the next several decades, land use issues continued as the principal objects of environmental attention. The interests involved were, accordingly, those of sportsmen and outdoorsmen—hunters, trappers, fishermen, hikers, bikers, and campers—as well as those of loggers, miners, and ranchers, who would profit financially from a Pinchot-formulated multiple-use policy reconciling preservation of the land's aesthetic and ecological values with carefully managed and controlled exploitation of its resources. Though bitterly fought, the politics of conservation were clearly delineated, with the sides sharply drawn and the issues relatively circumscribed.

But shortly after mid-century, things changed dramatically. The evolution from conservation to environmentalism brought with it a whole new complex of issues and concerns and a corresponding proliferation of advocacy groups. That evolution was set in motion in large part by one woman and one book.

The "New" Environmentalism

Rachel Carson's *Silent Spring* gave birth to what we now regard as the environmental movement and, with it, modern environmental politics. In it, Carson documented the lethal effects of chemical pesticides, particularly DDT, not just on their intended targets, but on all life, contaminating everything with which they come in contact and reaching humans by infiltrating the food chain as well as air and water supplies. Eventually, she warned, they threaten "the very nature of the world—the very nature of its life." By implication and extension, she attacked the abuse of science and the worship of technology, as well as the burgeoning post–World War II industrialism that was imposing increasing burdens both on natural resources and on public health. The substance of her claims, and the dramatic terms in which she cast them, captured the public's attention. The book went beyond description, however; it called for a new public consciousness of the dangers around us and for a more activist government to control them. In essence, it called for a new politics of environment.

As we now know, over the next several decades the public consciousness that Carson invoked did indeed develop, not only of the chemical pesticides that were her immediate target, but also of an ever-broadening list of domestic toxic threats. Industrial activities created ever-mounting and often uncontrolled repositories of solid and hazardous wastes; nuclear power plants generated radioactive waste stockpiles that society has yet to devise the means to sequester from humans; and oil spills have fouled previously pristine areas of our coastlines. People became concerned not only about the deterioration of their parks and forests but also about their more immediate surroundings. Henceforth they would be anxious about the purity of the water not only in oceans, lakes and streams, but also the water from their faucets. The quality of the ambient air—worsening under the onslaught of increased automobile traffic—remains an issue, but it has been joined by worry over indoor air pollution—"sick building syndrome" as it is called—as well as by radon infiltration of homes in some geographical regions. Microwave ovens and cell phones, sport utility vehicles and diesel trucks, decaffeinated coffee and food dyes, fertilizers and herbicides, lead paints and asbestos coatings, automobile batteries and used motor oil, antibacterial household detergents and antibiotic medications, plastic food wraps and dry-cleaning chemicals, and, ironically, alternatives to the very pesticides that were responsible for the movement in the first place—food irradiation and genetic modification. All these and countless others have been alleged to threaten public health and the environment and have, accordingly, been the subject of political wrangling and, in many cases, of legislation and regulation.

The identification of this growing matrix of potential risks raised a whole host of legal, ethical, and social questions, as well as concerns about potential liabilities and costs. Manufacturers, retailers, and governments clashed over the question of who was responsible for the safe management of these risks and whether appropriate warnings and notices were sufficient to discharge that responsibility. Many disclaimed responsibility entirely, and invoked the "buyer beware" principle. Employers clashed with workers over the safety of their workplaces, especially those that are the sites of industrial activities necessarily involving hazardous materials.

New land use questions arose, and these, too, took on a more local cast. The conservation of natural areas for their aesthetic qualities, as habitat, and as recreational resource is, as we have noted, historical. But recently, more thorny issues have come to the fore. Builders and developers have steadfastly resisted legal requirements to incorporate in their plans the impact of their construction on flood-prone areas, wetlands, habitats of endangered or threatened species, and fragile waterfront areas. All these challenge the extent to which government regulation can reach without treading on the constitutional right to private property. People became more sensitive, too, to the siting in their neighborhoods of, for example, unwanted solid waste, hazardous waste, and sewage treatment facilities that at worst exposed them to potentially higher levels of pollution or at least reduced the value of their homes and property. The NIMBY ("not in my backyard") syndrome became a familiar tag for these people. Many were opposed to any new development at all that increased traffic and air pollution and gave rise to the politically charged phenomenon of suburban sprawl.

Finally, civil rights and questions of ethics have become inextricably entangled with environmental activities and conditions. *Where* to site facilities that may expose surrounding populations is an environmental matter to be sure, but it has also become a civil rights matter, as such sitings have tended to follow the path of least political resistance, i.e., in economically disadvantaged and minority communities. Furthermore, national environmental issues have spread abroad, with the increasing exploitation by transnational corporations of impoverished third-world labor and the manufacture of products, the low cost of which reflects, in part, the environmental compromises made in the interests of economic competitiveness. More significant have been the bitter disputes over the commodification of third-world rain forests and other irreplaceable natural resources essential to maintaining biodiversity in response to the desperate basic needs of indigenous populations. Even animals have become an interest group. The production of cosmetics and fur coats, which involve what many consider the unethical treatment of animals, have now become hot-button issues and have garnered their own constituencies and created their own demands.

The transformation from conservation to environmentalism unleashed at mid-century has caused a transformation in environmental politics. In marrying public health with conservation, the new environmentalism has extended its reach to the whole population and made each and every citizen a stakeholder in the environment's well-being. By bringing the concerns "in from the outdoors," environmentalism has become a factor in our daily lives—in what we eat and how we grow it, where we live, what we buy, where we work, and even where we play. Witness, for example, the attention that environmental stresses associated with applications of pesticides, fertilizers, and herbicides required in golf course maintenance and the irresponsible dumping of solid wastes during leisure cruise ship operations are now commanding.

This fundamental change in the environmental agenda has, as one might expect, generated an explosion of new issues and new interests. No longer can one view the environmental wars as simply bilateral—government versus the private sector, environmentalists versus business, developers versus preservationists. The interests that have been created are much more numerous and complex and occupy many more narrow and focused niches. As the enumeration of policy positions on a range of issues recited earlier in the chapter suggests, they are also frequently unexpected and unorthodox. Now we have, as a particular issue may dictate, builders and developers of residential properties on opposite sides of the political table from industrial and commercial builders and developers; producers of high-sulfur coal are fighting off their low-sulfur coal counterparts (a battle that translates into a regional war) with respect to clean air regulatory regimes; and the national giants in any number of industries often separate themselves from the mom-and-pop operations in that same industry. Even wilderness hikers and rock climbers are doing battle, most recently over the prohibition imposed by the National Forest Service on fixed anchors on rocks, which climbers argue are conceptually indistinguishable from fixed markers on hiking trails. Finally, it must be recognized that the interests of state and local governments are different not only from one another, but from those of the federal government as well.

If the combinations and permutations of interests are almost endless, so too are the organizations, associations, alliances, interest groups, coalitions, and lobbies—permanent, temporary, or ad hoc—that have been created and entered into to influence public opinion and represent these interests in the formal policymaking process. Because the impact of legislative and regulatory policies can be subtle, yet significant and because counterinterests are all around (even where one may not expect them), each group feels it needs its own representation and its own public relations effort.

The changing role of environmental organizations is discussed in Chapter 7 in detail, but it might be appropriate here to hint at one of those changes with political ramifications. Some interests have less financial support and fewer members than their national counterparts. Their survival often depends on their linking up with other, sometimes only tangentially related, interests to maximize their political power and funding. Co-op America, for example, is a program to consolidate and direct the buying power of American consumers toward environmentally benign products, but it associates itself as well with those who are working more broadly for social justice—opposing sweatshops, promoting work for the disabled, and in general improving the life, skills, self-esteem, and dignity of men and women in low-income areas. Similarly, Natural Awakenings is representative of a growing number of local organizations that attempt to combine a group of personal health practices—acupuncture, chiropractic, massage, nutrition, and yoga—with "planetary health." Probably the fastest-growing new alliance is that between religious and environmental groups. Environmentalists and spiritual leaders are increasingly joining in efforts to redirect the environmental agenda by reawakening society's spiritual values and rethinking man's fundamental responsibility to nature.

It is generally acknowledged that U.S. environmental policy is the result of the combat and ultimate reconciliation of competing interests. It is the central premise of this book, however, that the number, nature, and impact of those interests have changed so dramatically over the past few decades that they constitute a new politics of the environment and that only by viewing environmental politics through the interest-group lens can one understand how we got where we are today with respect to environmental policy and, perhaps more important, where we are going and what forces will shape that destiny.

To be sure, the lawmaking process and the major combatants in the environmental wars in large measure remain in place and thus continue to merit our examination, but both have had to accommodate themselves to the changing environmental landscape. New legislative strategies have been devised and old ones revived to meet the new political scene, and the rulemaking protocol has undergone a virtual metamorphosis. Both the legislative and regulatory processes, at least in practice, now upset many of our civics class notions of how they operate. The catalysts and impact of these changes on environmental policy are the subjects of Chapters 2 and 3, respectively.

These changes in the policymaking process and in the composition and strategies of the major participants in that process represent significant political change by themselves, but the changes go beyond refurbished theaters and old actors in new roles. New participants with deep influence have emerged from the newly defined environmentalism. The most significant of these are the media. What I call the "domestication" of the environmental agenda has arguably made the media the most important players in environmental politics today. It is not a role that the media either sought or willingly accepted. But as environmental issues and concerns centered increasingly on our homes and workplaces and food and lifestyles, newspapers and television found much more in environmentalism that directly engaged their audiences, though in the minds of many environmentalists, that has been a mixed blessing. The issues that have been given the most prominent treatment are by no means the most significant or threatening. Nevertheless, by some "invisible hand" (to borrow Adam Smith's famous phrase), public concerns do get communicated to official policymakers. For that reason, interest groups from across the spectrum have tried to enlist the media in their efforts to set the agenda and accomplish their purposes. How the role of the media in environmental politics has evolved and how it has affected public policy are examined in Chapter 4.

Chapter 5 takes up another newly enlisted participant in the environmental wars—the scientific community. Another legacy of the Carson-inspired movement is the bringing of public health concerns under the environmental umbrella, a legacy that puts the grounding of much environmental policy beyond the reach of laymen and thereby creates a substantial role for scientific experts. Despite the reticence of scientists to provide the superficial answers to complicated problems that policymakers seek, interest groups of all types have eagerly sought them out and capitalized on the credibility with the public they can bring to their positions. The resulting politicization of science is here examined in detail.

State and local governments have themselves become interest groups in response to the new environmentalism and thus cannot simply be subsumed under the broad "government" rubric. Why this has happened, and its effect on their traditional responsibility to implement environmental law, is addressed in Chapter 6.

In the minds of most people, the real environmental war is between big business, whose industrial and commercial activities have imposed the most strain on the environment, and the environmental establishment, which has served as agenda setter and self-appointed watchdog of government's regulation of the private sector. Seen in the broadest terms, that, of course, continues to be true. But the environmental establishment and the business community have themselves undergone significant reinvention in response to the growing influence of the newly emergent complex of environmental interests. Each has been criticized by the purists in its own camp, but the reality is that they have come much closer to each other and have even worked together on specific programs for practical, if not ideological, reasons. Further, each has sprouted new "branches" that operate independently but serve many of their same ends. Environmental justice groups and the more controversial ecoterrorists supplement, but by no means mirror, the efforts of more mainstream environmental organizations. Similarly, corporate-funded conservative and libertarian think tanks and a consortium of interests constituting the Wise Use movement try to emasculate the government's environmental work and keep environmental organizations in tow. They strengthen the antienvironmental forces in the halls of government and in the popular media but allow the business establishment to pursue the same goals in more politically conventional ways. To complicate matters even more, organized labor has played a somewhat paradoxical role, siding with business interests when jobs are at stake and siding with environmentalists when business practices jeopardize their well-being. These two camps and their respective evolutions and allies are examined in Chapters 7 and 8, respectively.

Two other recent developments promise to influence, if not fundamentally alter, the politics as well as the substance of environmentalism in the years ahead. It may seem inconsistent, if not contradictory, to assert that even as local environmental threats are displacing national ones as the primary subjects of policy proposals, environmental conditions and concerns that reach beyond our borders are imposing themselves on U.S. environmental policy, but that is indeed the case. It is beyond the scope of this book to examine the environmental policies and politics of the rest of the world. It is important to realize, however, that international affairs and attitudes are playing a growing role in shaping our national policies on many issues. The nation's air and energy policies, for example, are at the center of the debate over whether we should expeditiously implement the Kyoto treaty, signed by the U.S. and scores of other nations toward the end of the last century, to address global climate change. An even more dramatic example is the international community's harsh response to genetically modified foods, which not only is causing a reevaluation of U.S. agricultural practices but also is the catalyst for a political battle among the biotech community, farmers, food processors, food marketers, and a host of other interests that are affected in one way or another. The political implications of these and other effects of globalization and free trade are discussed in Chapter 9.

Finally, and ultimately most important, is the revolution in information and communications technology of the Internet. It is the most important because, without it, so much of what is the new politics of environmentalism could not have taken place, or at least would have played out quite differently. In making possible the inclusion of, and communication among, the myriad groups that are joining in the environmental wars, the Internet has become the "great facilitator." Its political ramifications for environmental policy today and in the future are the subject of this book's concluding chapter.

The broad and systemic evolution of environmentalism made it the most significant social and cultural movement in the second half of the twentieth century. It has also been the most profound and interesting political movement. Almost no other development has commanded as much public involvement, principally because no other has had such an impact on people's daily lives. To track environmental politics in action, one does not have to visit the halls of Congress or state capitols. One need only open the daily newspaper, watch television news reports, pay more attention to junk mail, or log on to the Internet, where untold numbers of interest groups vie for public attention and thereby for political favor. Students of environmental politics are encouraged to do this as their formal course work progresses. It is hoped that this book will help guide that effort.

CENTRAL IDEAS

Environmental politics is driven by interests rather than by ideology. Over the past several decades, the number and nature of interests have changed dramatically, transforming environmental politics from contests waged by powerful interests over conservation to local wars over public health. The Internet has played a key role in facilitating this transformation.

Chapter 2

Legislation: Leveling the Playing Field and Leveraging the Process

> If you like laws and sausages, you should never watch either being made.
>
> Otto von Bismarck

> Discovering a workable definition of environmental law is a little bit like the search for truth: the closer you get, the more elusive it becomes.
>
> William H. Rogers

The formal procedure by which laws are enacted at the federal level and by the states is designed to reflect the most cherished values of democratic society—openness, equity, inclusiveness, and stability. It is a process that is deliberative, structured, and predictable and, perhaps most important, that provides the citizens of this country with the comfort—justified or not— that power is shared and that everyone participates, either directly or through representatives. Americans assert with pride that they are "a nation of laws, not men."

At the same time, and paradoxically, Americans knowingly smile at Chancellor von Bismarck's famously cynical characterization of the making of laws cited above, because the realities of the legislative process are more complex than might be expected from a superficial analysis of the bare-bones procedure itself. Simple common sense would inevitably lead us to the fundamental truth that those in positions of authority and wealth, and those with the greatest stake in its results, exercise disproportionate power. The statutes, therefore, clear and compelling as they may appear, belie the extraordinary mix of influences that go into their formulation.

Because environmental issues affect us in such profound ways and involve such a multiplicity of interests, they subject the legislative process to its greatest strains. Over the course of the last quarter century or so, the body of laws enacted in the environmental area have influenced almost every aspect of our daily lives and mandated the appropriation of more public and private monies than the body of laws in any other area. Perhaps more significantly, environmentalism and politics are inextricably linked because substantive values are at stake, to say nothing of behavior. Every major policy supports some interested party's view of the world and adversely affects someone else's. That is why the collective process for solving problems related to the environment is inevitably adversarial. The legislative process tries to minimize, or at least manage, these divisions, provide the circumstance for the reconciliation of conflicts, and allow problems to be thoughtfully and rationally resolved. And by and large it does. But at virtually every step in that process are entry points for interests to promote their causes or opportunities to stifle threats to those causes, and that is why, despite the best efforts of our framers, politics finds its way into lawmaking.

To appreciate environmental politics as it is currently practiced, a brief perspective on its history might be useful. Until about the mid-twentieth century, environmental threats to private property and public health were usually handled by the courts in accordance with common law principles and, in certain limited areas, by state and local governments. Under common law, the burden was on those who called for action to demonstrate how a specific party engaged in acts or created a circumstance that specifically and unreasonably caused environmental harm to them. A plaintiff had to suffer personal damages before being able to bring an action against a polluter or developer. "Public injuries" were pretty much a foreign notion, and economic interests were given preference over social needs in controversies over land use.

All that dramatically changed in the latter half of the twentieth century. The post–World War II economic boom, in the absence of any significant governmental restraints or oversight, imposed an increasing burden on the nation's resources. Pollutants created by industrial activities were routinely discharged into the air and bodies of water; solid wastes were dumped into public landfills, along with any number of toxic wastes; and pesticides developed from a wide range of untested chemicals were often indiscriminately applied in agricultural and horticultural production. The nation's reserves of wood, oil, and minerals were extracted with little restraint, and wetlands and habitats of endangered species were claimed by rampant development. By mid-century, some of the signs of these profligate activities had begun to manifest themselves.

But the modest initial efforts of the federal government to address these problems did not, at first, generate private-sector resistance. Two measures—the Federal Water Pollution Control Act (1948) and the Air Pollution Control Act of 1955—were broad, general approaches to their respective problems, and the robust economy fueled by the postwar economic boom subsumed most of the regulatory costs imposed by these measures. Further, few interest groups at this time had sufficient expertise and research capacity to get deeply involved in the lawmaking process. It was not until later, when it became apparent that the goals of these statutes were not being achieved, that their more rigorous reauthorized versions began to command more attention.

Rachel Carson's *Silent Spring* (1962), as noted in Chapter 1, ushered in the modern environmental movement and, in so doing, gave birth to contemporary environmental politics. Its allegations against the chemical industry, in arousing deep public concern over the health effects of broad pesticide applications and by stimulating sharp debate over the proper role of government in addressing this threat, became a paradigm for subsequent battles over endangered species, toxic and solid waste reduction and disposal, "Superfund" site cleanup, marine protection, nuclear power generation, and a host of other issues. From that point on, environmental policy would be in the public consciousness, and both the media and many new local environmental watchdogs would join the newly vigilant national organizations to set an aggressive agenda and begin to shape federal action to address public concerns.

After the broad, comprehensive framework to address the nation's overarching environmental problems was set in place, more narrowly focused measures to address specific pollutants in air or water supplies, special classes of toxins, threatened or endangered species, ecologically fragile areas such as wetlands and coastal areas, and a host of other newly emerging potential health threats such as lead, electromagnetic fields, and food additives forced their way onto the scene. As the scope of the problems narrowed, so environmental issues inched closer to home. Concerns over ocean and lake quality were joined by fears over the quality of water from home faucets. The quality of ambient air competed for attention with anxiety over indoor air quality, particularly the presence of radon, and the environmental safety of our workplaces, schools, and food supplies commanded increasing legislative attention.

This "domestication" of the environmental agenda was accompanied by two other factors that gave rise to a more activist political climate. As Chapter 3 details, the EPA exercised its growing authority by tightening regulatory standards across the board as the capacity to detect and remove increasingly microscopic concentrations of pollutants improved, inevitably increasing the cost of compliance to the business community. But the private sector was, at least until the late 1980s, on the defensive as a result of a series of highly publicized, galvanizing events that aroused the public: the oil spill off the Santa Barbara coast, the Love Canal and Times Beach episodes, the nuclear accidents at Chernobyl and Three Mile Island, the Bhopal disaster at an American chemical plant in India, and the *Exxon Valdez* oil spill. These environmental disasters or near disasters invigorated the environmental movement, generated increased membership and funding, and virtually earned it the proxy of the American public. Those feeling the pinch of the activist legislative agenda had to turn to the less visible regulatory arena for relief.

Environmental statutes, then, have historically taken on the cast of public problem solvers and the protectors of the public welfare. Their missions—embodied in their introductory "Declarations of Goals and Policy" sections—establish laudable objectives to be achieved by their operative provisions, either promoting the public welfare or, even more compelling, preventing potentially catastrophic threats to public health and safety or the ecosystem. As such, they are hard to impeach, and they seldom are. Rather, those who would escape their reach are forced to adopt strategies that minimize their effect on them or somehow escape their purview altogether. These strategies can be best understood in light of the legislative process itself, to which we now turn.

The process starts formally with the introduction of a legislative bill, which is simply a proposal for a new law or for amendment of an existing one. Such bills are sponsored by a member of Congress, either a Representative or a Senator, and entered into the agenda of the House of Representatives or the Senate. After a bill is introduced, it is referred to a committee or committees. The Congress obviously cannot collectively consider each introduced bill, which would be far too cumbersome, complicated, and confusing, to say nothing of time-consuming. So Congress has organized itself into what are referred to as "standing reference committees," each with a particular subject matter, on which its members presumably have some expertise. It is not uncommon for more than one committee to consider a bill and mandatory that the appropriations committee be one of those when the bill involves substantial public expenditures. After a bill is considered, discussed, and amended if deemed appropriate by the members of the committee, it is "reported" by a majority vote of the members. Once an agreed-upon version of the bill clears all relevant committees, it is sent to the full House for a vote. A majority vote passes the bill, except in the special circumstances noted later.

After a favorable vote by the full House, a bill moves to the other congressional body, where it follows essentially the same procedure. If identical bills are introduced in both the House and the Senate, the bills can, of course, move sequentially or concurrently. In both cases, identical versions of the bills must pass both houses. In state legislatures, bills usually continue to bounce back and forth until the same version can be mutually agreed upon. At the federal level, there are what are called "conferences," which are meetings between representatives of the House and Senate to reconcile major differences between the two versions of a bill and prepare a "conference report." Unlike virtually all other legislative gatherings, conferences usually are not conducted in public. Minor differences between the House and Senate versions of bills are usually worked out by concurrence.

When agreement on a bill has been reached and a majority of all reference committees and both Houses have voted on it favorably, it is sent to the president for final approval and passage into law. The president can veto it, conditionally veto it with recommended changes, or sign it. A presidential or gubernatorial veto can be overridden by a larger majority of the full House, generally two thirds, but conditional vetoes are usually either accepted with the recommended changes or negotiated with the administration.

This, in skeletal form, is the trip that a bill takes to become law. It should be obvious, however, that, despite the superficial objectivity of the process, many political and policy implications are submerged in it. At virtually every stage, interests can penetrate the procedure and move, stall, or

How Laws Grow

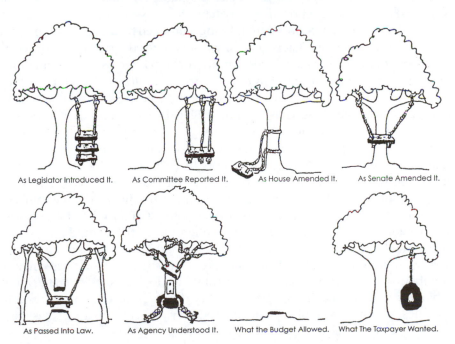

As Legislator Introduced It. As Committee Reported It. As House Amended It. As Senate Amended It.

As Passed Into Law. As Agency Understood It. What the Budget Allowed. What The Taxpayer Wanted.

amend a bill's provisions to suit their purposes. These opportunities merit closer attention.

Since the process starts with posting a bill on the legislative agenda, that is the first point of influence. Bills come from a variety of sources. It must be said that the legislation banning DDT spurred by Rachel Carson's *Silent Spring* is not typical. It is rare that a book achieves such public notice that it drives public policy, but in the social and political milieu of the 1960s, it happened. It was not until 30 years later that another book, *Our Stolen Future*, captured considerable attention and served as the catalyst for a major law, The Food Quality Protection Act of 1996. *Our Stolen Future* alleged that a class of chemicals known as endocrine disrupters were potentially responsible for a number of adverse health consequences, including reduced sperm counts (hence its title). In both thesis and reception, it bears an uncanny resemblance to its estimable predecessor.

Still, books are rarely political catalysts. Somewhat more commonly, though not typically, major environmental incidents such as those mentioned earlier stimulate legislative action. Though such events do pull together all the forces needed to move legislative responses decisively, they fortunately do not occur with sufficient frequency to determine policy, and environmental advocates cannot very well hope, even privately, for new environmental calamities to renew their mandate. When an environmental circumstance, such as a drought or flood, has the potential to drive policy, the media can often turn it into a political issue. But most often, less spectacular issues are brought to the legislative table by interest groups of all kinds, an investigative report they may have commissioned, a court decision inviting legislative direction, or their being aggrieved by the implementation of existing laws, though that is generally reserved for the regulatory process.

Interest groups that have sufficient influence with legislators sometimes go beyond the active promotion of a measure and literally produce a draft for introduction, a practice that is no doubt rather more common than generally acknowledged but that is essentially illegal. Two examples of improper interest group involvement in the very earliest stage of the legislative process—the preparation of legislative bills—gained notoriety from press exposure in the mid-1990s and early 2000. In 1995, a prominent wood products company drafted key provisions of a bill that would have insulated alleged violators of law from being prosecuted by the EPA in certain circumstances (among which was theirs). In June 2000, a pesticide lobbyist openly admitted to crafting a bill for a congressman that would have effectively blocked implementation of the same Food Quality Protection Act of 1996 that derived from *Our Stolen Future*. Only slightly less activist are any number of firms surrounding the Capitol that offer "legislative strategy and advocacy services." Such firms usually carry out traditional lobbying responsibilities, but in providing such advertised services as "preparing legislative materials" and "developing legislative strategies," they can penetrate and skew the legislative process by contacting key legislators before the bill is formally considered. Certainly, the ability of well-financed, politically connected interests to get a measure on the agenda is infinitely greater than that of a lesser group, particularly at the national level.

Another element of politics in the process resides in the officers themselves. The Houses have leaders—the Speaker in the House of Representatives and the President in the Senate. They have many sources of influence, from the selection of committee chairs to the control over the agenda they maintain by committee references to the allocation of monies in political action committee funds (PACs) to which lobbyists contribute. Either by themselves or through their chairs, they can kill bills or they can accelerate their movement through the successive stages. Committee chairs also have a measure of control. They are under no obligation whatever to even consider any particular bill referred to their committee. There are vastly more bills than can ever be considered, and only a small percentage of them ever get on a committee agenda, much less voted on. Bills can, of course, be heard and tabled, in effect killing them. By either tabling them or failing to take them up at all, committee chairs can simply bury bills they don't like or that their more influential "constituents" don't like. Speaking of the "influence" and "access" of politically and economically powerful interests, then, refers not only to public votes but also to almost invisible control of the legislative agenda through its presiding officers and chairs.

Further, committee members themselves sometimes use the bill review process to advance the specific interests of their constituents, even on measures unrelated to that of the bill. Technically, of course, they are *United States* Senators and *United States* Representatives, who are formally committed to protecting and promoting the welfare of the nation. But they are, of course, elected by the voters in their respective states and expected to represent their local interests as well. The principal constituent concerns of a senator from, say, New York are in most cases radically different from those of a senator from Iowa or Montana. Thus it may well—and often does—serve the interests of all concerned to "trade" votes on proposals before them. By supporting a measure important to a fellow committee member, or House member, in exchange for his or her vote on an issue important to constituents, a legislator can secure something that is significant to his electorate while giving up little in return. Thus are majorities among diverse bodies achieved.

Government agencies can become lobbyists as well. The first committee action taken on a bill is usually a request for comment by the government agency or agencies with regulatory jurisdiction over the subject matter, and they have their interests, too. Environmental regulatory agencies such as the EPA, the Department of Interior, the Fish and Wildlife Service, and the

Bureau of Land Management have institutional interests in what policies emerge and the shape they take, because their authority and budgets are potentially at stake. Aside from the considerable policymaking influence they exert in adopting regulations, they have an interest in the laws themselves, which give or deny them the authority and funding to carry out the provisions of the statute. It would be naïve to assume that these agencies would actively pursue the general public interest without thinking at all of their own institutional interests. "Bureaucrats" are often accused of "careerism" when they promote policy initiatives. Although it is their alleged regulatory excesses that are at the source of these critiques, their very real influence on legislation is often overlooked.

The committee consideration process offers many other opportunities for the input of interest groups that have the time, expertise, and money to stay abreast of the congressional calendar. Because, as noted earlier, the titles and missions of major bills promise to be all things to all people, changes to the introduced draft are offered during committee meetings, which are sparsely attended and, by and large, out of the public eye. Still, requests for changes advanced by interest groups are most often technical in nature so as not to appear to oppose the lofty objectives of the proposal. As generally well informed as committee members are, they can seldom challenge the data or rationale provided by experienced lobbyists steeped in the particular subject matter. Legislative staffers are the principal guides for committee members on detailed matters, but they cannot pretend to the expertise of the private sector's legislative agents, so that, by the time a bill wends it way through two or three committees in each house, it is most often different from the introduced draft in significant ways. Partisan, geographic, and interest group considerations have been forced on it at every stage, and although the reported bill may have the same title and public purpose as the original one, it often shows the scars of the legislative wars.

It should be apparent, however, that the virtues of this step-by-step, open, and deliberative process are also its limitations. They render the legislative process dispositive toward certain kinds of policy outcomes and resistant to others, except, of course, when dealing with a highly charged emergency. Because we prefer incremental change to dramatic change, only some of the potential alternatives for dealing with a problem are considered, which are generally the most practical and readily implemented ones. Limitations on the amount of information available to policymakers and on their time, together with the need for consensus, preclude comprehensive change. Thus for many practical and political reasons, bills advance by easy stages. Only crises can ring in dramatic new laws.

Along with the disposition to modest, incremental change is a short-term policy bias. The 2-, 4-, and 6-year terms of office to which legislators are elected, and the fact that most budgets are required to be submitted annually, inevitably dispose lawmakers to policies that show immediate results and reduce immediate costs. This disposition is especially harsh on environmental problems, which do not generally lend themselves to quick fixes. Gross environmental and ecological conditions take time to address effectively, and initial investments often do not demonstrate a return for long periods of time. Again, spills or accidents that constitute crises can respond in the short-term to initiatives and quick infusions of money, but most environmental problems swim against the legislative current. Perhaps this is why crises, or the appearance of them, are so often exploited by environmental groups and why their dramatic presentation in the media has been such a potent driver of legislative action.

The decentralization of power in the legislative process that protects the public from autocratic rule also works against prompt, constructive action. An axiom in the halls of government is that it is much easier to keep something from happening than it is to make something happen. If the many and varied interests playing into the bill review process offset each other or get caught up in partisan wrangling, legislative gridlock results. When these circumstances obtain—as they often do during the final months of an administration—ways to get around the plodding, Byzantine legislative process are sought. It was not hard to find such conditions during the Clinton administration, especially during his second term.

The systematic and deliberative procedure by which legislation is formally enacted notwithstanding, especially contentious issues—and especially contentious times—often call for extraordinary strategic maneuvers. The 1990s were such times, and, accordingly, Congress and the President revived and expanded tactics that enabled them to circumvent or co-opt one another to pursue goals that they almost surely could not achieve through the conventional process. I am speaking of riders and executive proclamations, respectively. Let me stress clearly at the outset that both of these tactics are perfectly legal, and each has been used throughout our nation's legislative history. But recourse to these tactics in the area of the environment increased dramatically in the 1990s, during which the Executive and Legislative branches were for the most part controlled by different parties and were as bitterly at odds as they have been for years. Let's take a closer look at them, for they have become much stronger forces in shaping environmental policy today.

Legislative Riders

Riders, provisions added to bills not related to their main purposes, have been with us almost since the beginning and have been responsible for countless measures that have become part of our established law. Unlike the constitutional authority vested in all but a half dozen or so state legislatures, the federal Constitution permits provisions about issues unrelated to the purpose of the original bill to be considered and voted on together with the original bill by the lawmaking body. Thus a special-interest group may prevail on lawmakers to tack a controversial provision it desires onto a bill, usually one that is popular and likely to pass, which "carries" it to passage (thus the nickname). In this way, riders escape the scrutiny of committee hearings and floor debate built into the legislative process.

The most common vehicles for riders have historically been appropriations and spending bills, because both types of bills inherently have a diversity of purposes. In addition appropriations bills often have an urgency about them, particularly those dedicated to relieving some grievous situation and budget legislation is shielded from filibusters in the Senate, thus minimizing debate and opposition. Two enacted riders illustrate how substantively unrelated riders can be from the bills to which they are attached and how they can advance on the merits of their "carriers." A rider reducing the U.S. Fish and Wildlife Service's budget for listing endangered species became law on the back of the spending bill funding the peacekeeping mission in Haiti. And a number of antienvironmental riders were attached to the Kosovo emergency funding bill, though most were negotiated out of the final measure. Riders have historically served legislators well as means to secure funding for their own districts' projects, otherwise known as pork barrel projects. Although no doubt responsible for draining the national treasury in the service of special interests, they have, by and large, historically not played a major role in policymaking.

This approach changed significantly in the 1990s, especially in the area of environmental policy. The change can be traced back to the election of the 104th Congress and Speaker Newt Gingrich. Despite the best efforts of previous administrations and the rising influence of antiregulatory forces since 1980, virtually all polls confirmed that the public remained strongly committed to environmental protection, whatever its costs. Environmental activism, however, did not comport well with the platform of congressional Republicans and many conservative Democrats. The unveiling of an old tactic, the rider, seemed like an ideal way to advance their agenda with minimal political fallout, especially because the antienvironmental initiatives subtly embedded in the Contract With America failed to secure Senate passage and largely fell by the wayside.

President Bill Clinton first encountered this tactic when he made an active effort to resolve the issue concerning logging in the Pacific Northwest, while trying to find a meaningful compromise between those who wanted to protect the resource and the economic interests promoting continued logging. The environmental forces appeared to have won with the Fiscal 1995 Supplemental Appropriations and Rescissions bill, which proposed more

Dunagin's People

"This bill is strong enough to carry
a dozen senseless riders!"

than $17 billion in spending cancellations for budget accounts approved by the preceding Congress. Representative Charles Taylor, however, attached a rider to it to increase logging on U.S. Forest Service and Bureau of Land Management lands. The rider provided for the suspension of all federal environmental laws to log the ancient forests of the Pacific Northwest and to "salvage" log at least 6.2 billion board feet of trees affected by wildfire or insect infestation. Although President Clinton did veto this particular bill, he signed a similar one 90 days later. The game was on.

Expectedly, President Clinton was faced with a series of fiscal 1996 appropriations bills laden with antienvironmental riders. These measures were by no means purely budgetary; they contained significant policy implications. Among them were measures to eliminate EPA's role in protecting wetlands, to restrict energy-conservation and -efficiency programs, to impose a moratorium on listing for protection any additional threatened and endangered species, to establish a timber plan permitting considerable logging of the Tongass National Forest in Alaska, and to reduce the scope of the California Desert Protection Act by relieving the National Park Service of the responsibility for managing the Mohave National Preserve. In addition, an omnibus bill in the House of Representatives consolidating the proposals would have cut funding for environmental programs and reduced EPA's budget by 10%.

President Clinton's threat to veto the entire measure ultimately saved most of the programs and the EPA budget, but provisions to halt endangered species designations and expediting the logging of ancient forests were signed into law as part of emergency spending legislation. This was only the beginning of what was to become a continuing assault on environmental protection through an essentially clandestine legislative strategy.

In 1998, more than 40 antienvironmental riders were attached to appropriations bills for the departments of the Interior, Transportation, and Commerce. In an unsuccessful attempt to forestall this stealth strategy, Representative Henry Waxman introduced the Defense of the Environment Act, a bill that would require open floor debate and a separate vote on all proposals that would have the effect of weakening environmental protection. Meanwhile, the earlier practice continued. The Alaska delegation, the members of which hold influential seats on appropriations and resource committees, advanced a number of riders to spending bills designed to facilitate exploitation of Alaska's natural resources. Further, riders to the Veterans Affairs—Housing and Urban Development appropriations bill that would have also delayed cleanup of toxic mercury and PCBs, delayed federal action to protect children from harmful pesticides, and interfered with EPA efforts to clean the air in national parks were also introduced.

In 1999 more than 50 antienvironmental riders were attached to bills funding the budgets of the departments of Commerce, State, Interior, Defense, and Agriculture and the EPA. Again, the issues were more than fiscal, ranging from climate change to mining waste, to wildlife protection, to royalties for extracting oil from public lands, to grazing on public lands. In another defensive strategy, Representative Norman D. Dicks successfully advocated a directive to House negotiators on a measure funding the Department of the Interior to reject Senate riders. Though of little substantive influence, it was one of a number of elements that collectively led to the ultimate negotiation of a bill without the riders.

To demonstrate that Republicans are not the sole practitioners of legislation by rider and that interests prevail over partisanship, Senator Robert Byrd of West Virginia, the senior Democrat on the Appropriations Committee, sponsored a rider that would have exempted mountaintop coal mining practices in his state, a measure opposed by House Democratic leaders as well as a group of House Republicans. If enacted, the provision would have effectively overturned a federal court decision finding that West Virginia's mining companies were in violation of clean water and surface mining laws.

The result of rider-wrangling in the 105th Congress was, by and large, won by environmental interests. Only a rider preventing the Department of Transportation from raising fuel economy standards for cars and light trucks was subsequently enacted. Riders proliferated in Fiscal 2000 appropriations bills, and continued efforts to obstruct compliance with U.S. commitments under the Kyoto climate change treaty, to forestall fuel efficiency standards (known as "corporate average fuel economy," or CAFE, standards), to promote logging of forests, and to minimize or eliminate reviews by regulatory agencies are embedded in them. In fact, the Natural Resources Defense Council, at a link on its website (nrdc.org/legislation/crossroads/chap4.asp) identifies 70 Fiscal 2000 budget riders that were modified and defeated. But even when riders fail to be enacted, their goals are not infrequently embodied even less visibly in the "report language" through which congressional committees express their "legislative intent." Such language, though technically not "the law," must be taken seriously by regulatory agencies.

So pervasive and potentially threatening have antienvironmental riders become that at least three environmental organizations—the League for Conservation Voters, the Natural Resources Defense Council, and the Defenders of Wildlife—have established separate website links for the sole purpose of tracking and scorecarding these riders. That most such riders have failed to be enacted is attributable, in large part, to the increased exposure to which the rider strategy has been subject in the last several Congresses. At a minimum, antienvironmental riders have strained the resources of environmental activists, gotten measures onto legislative agendas that could not have made it on their own, and not infrequently forced compromises that would otherwise not have been made. And, as has been shown, in several areas—fuel efficiency standards, climate change, and commercial logging—they have virtually become established policy.

As Congress has sought to advance issues and interests through recourse to riders when those issues and interests could not survive the open legislative process, so has the executive branch engaged in its own extralegislative tactics to accomplish purposes it probably could not have gotten Congress to agree to. In fact, on July 5, 2000, the day after the nation celebrated its 224th birthday, the *New York Times* devoted facing pages to two related executive branch exercises of singular power. The first described the battle raging over the proposal of the White House to preclude, by regulation, road-building on 43 million acres of national forest land and thereby impede further development on as much as one quarter of the nation's entire forest system. The National Forest Service was holding extensive hearings on the proposal, but the ultimate power remains in the executive branch.

The other story more generally described the deliberate efforts of the president to rule by "decree" whenever he thought that he could not accomplish his goals by process. It publicized the various tools available—executive orders, memoranda, proclamations, and regulations—that he had used to move initiatives that otherwise would have stood little chance of passage by Congress. This is no mere sideshow. The Office of Management and Budget estimates that President Clinton averaged one executive order a week during his terms in office. Policymaking by the executive alone has a long and hallowed history, but President Clinton has perhaps exploited it more fully than any of his predecessors.

More stringent air pollution standards have secured wide publicity and engendered much nationwide controversy, but the most actively exploited and controversial initiative that the executive branch undertook at the turn of the twenty-first century is the protection of special natural and wild areas from commercial exploitation and development via another tack. Sensing the relentless efforts of Congress, through both legislative and court challenges to land use regulations based on Constitutional "takings" principles, President Clinton exploited the authority in the Antiquities Act of 1906, a law that allows the president to act unilaterally. The law states: "The President of the United States is authorized, in his discretion, to declare by public proclamation historic landmarks, historic and prehistoric structures, and other objects of scientific interest that are situated upon the lands owned or controlled by the Government of the United States."

There are, as a close reading of the law reveals, subtle but real differences between what can be designated a "monument" and what can be preserved as simply "wilderness" or natural lands. Treasures that can be protected by presidential proclamation are special, particular resources rather than whole ecosystems. Nevertheless, President Clinton saw in the authority vested in him by the Antiquities Act of 1906 a chance to build a preservation legacy— and endear himself to a significant component of the environmental community—by setting aside vast areas of federal lands without, and even in the face of, an angry Congress.

President Clinton's first significant act under this authority was the designation of 1.7 million acres of red rock cliffs in southern Utah as the Grand Staircase–Escalante National Monument in September 1996. He sought to enhance the significance of the designation and its symbolism by signing the proclamation virtually at the site where President Theodore Roosevelt proclaimed the Grand Canyon a monument in 1908 and thus immunized it from commercial exploitation.

While the President's act was, of course, lauded by preservationist groups such as the Sierra Club and the Southern Utah Wilderness Alliance, as well as the region's celebrity environmentalist Robert Redford, it was bitterly denounced by the entire Utah congressional delegation as well as the lone Democrat in the Utah State Legislature. They lamented the adverse effect of the designation on what is regarded as the largest known coal reserves in the U.S. Two large mining concerns were planning to undertake operations forestalled by the act. Andalex Resources and Pacific Corporation, both of which hold federal mining leases in the area, highlighted the crippling effect it would have on the economy and jobs. The Utah Association of Local Governments was also outraged, principally because mining royalties are dedicated to funding public schools, further exacerbating the political fallout of the declaration. Joining the chorus of angry cries were those of the Western States Coalition, county government agencies, and citizen groups. All pledged to fight the designations with every weapon in their power, but most admitted that they were not optimistic about their chances.

Most relevant here is that those opposed to the designation denounced the manner in which it was made as much as the action itself. While Senator Orrin Hatch conceded that the 1906 Antiquities Act gives the president very broad powers, and that blocking this designation would be extremely difficult, he and others who were offended by the exercise of what all agree was within legal limits used the most extreme language to attack its inherent unilateral authority. Representative Enid Greene said, "This is just not how a Democracy is supposed to work." Senator Larry Craig called it "a misuse of power . . . one of the cutest political tricks that I have seen in a long time. . . . To operate in the dark of night outside what is now believed to be the law is something that this president will have to be questioned on." Others in lesser public positions had even stronger opinions. A local mayor stated, "The Constitution was not written up for one man to have that much power," and a local resident felt "a little betrayed, like I'm in a communist country."

Clearly, President Clinton's act excluded a wide range of interests—mining, tourism, and education to name a few—from the public debate. And the vehemence of the reaction signals the difficulty the president would have had in pursuing his objective through the open legislative process. That was the rationale for invoking the Antiquities Act of 1906. His action did provoke more than a verbal response, just as the proliferation of riders spawned "defensive" legislative measures, however fruitless. First was a lawsuit brought by the Western States Coalition in November 1996, which charged that Clinton's decree violated the Antiquities Act of 1906 by exceeding its scope. The suit claimed that presidential action taken under the act should "be confined to the smallest area possible." In February 1998, a joint resolution was passed in the Alaska legislature that would require congressional consent before withdrawals under the Antiquities Act could be made. Concurrently, bills sponsored by Senators Hatch and Murkowski were being considered. A year later, Congressman Hansen of Utah introduced a bill to weaken presidential authority under the act, specifically providing for public participation in the declaration of national monuments under this authority and for their review under the National Environmental Policy Act. Two years earlier, Congressman Hansen had introduced a measure to require congressional as well as gubernatorial and state legislative approval of designations in excess of 5,000 acres. In sum, opponents of Clinton's action claimed

PRESIDENT CLINTON'S NATIONAL MONUMENTS LEGACY

MONUMENT	STATE	ACREAGE	
Grand Canyon-Parashant	AZ	1,014,000	
Aqua-Fria	AZ	71,000	
California Coastal	CA		8,000 small islands
Expanded Pinnacles	CA	7,900	
Hanford Reach	WA	197,000	
Cascade-Siskiyou	OR	52,000	
Ironwood Forest	AZ	135,000	
Canyons of the Ancients	CO	164,000	
Giant Sequoia	CA	328,000	
Anderson Cottage	DC	2	
Grand Staircase-Escalante	UT	1,700,000	
PREVIOUS TOTAL		**3,668,902**	**Plus 8,000 small islands**
Santa Rosa and San Jacinto	CA	281,600	
Vermilion Cliffs	AZ	293,000	
Expanded the Craters of the Moon	ID	661,000	
Upper Missouri River Breaks (Fed Land)	MT	377,346	
Upper Missouri River Breaks (River)	MT		149 miles of river
Pompeys Pillar	MT	51	
Carrizo Plain	CA	204,107	
Sonoran Desert	AZ	486,149	
Kasha-Katuwe Tent Rocks	NM	4,148	
Minidoka Internment	ID	73	
Expanded U.S. Virgin Islands Coral Reef	VI	12,708	
Expanded the Buck Island Reef	VI	18,135	
NEW TOTAL		**2,338,317**	**Plus 149 miles of river**
GRAND TOTAL		**6,007,219**	**Plus 8,000 small islands and 149 miles of river**

Reprinted with the permission of the League of Conservation Voters

both that he had exceeded his authority and that the act should be changed to require broader participation in such decisions. None of these challenges had much of a chance at passage, but they did direct the anger and frustration of opponents into formal channels of opposition.

These various formal initiatives did not discourage President Clinton from continuing the practice. In early January 2000, utilizing the same authority, he designated three additional monuments and expanded a fourth. The designations were consistent with recommendations developed by Interior Secretary Bruce Babbitt in 1998 for preserving unique and fragile places. These additions brought the total acreage newly set aside by Clinton as monuments to 2.7 million acres, with his expressed intention to subsequently add another 6 million. In April, Secretary of Agriculture Glickman recommended that 328,000 acres of national forest, housing half the giant sequoia groves still extant in the U.S., be protected, again under the authority of the Antiquities Act of 1906. The President allowed his Vice President (not incidentally running for president himself) to announce the designation of the Cascade-

Siskiyou and Hanford Reach National Monuments shortly thereafter. As they had before, opponents sought a legislative remedy. Representative William Thomas and two of his California colleagues introduced a bill requiring a study by the National Academy of Sciences as a precondition to designation. Unilateral action, however, can invariably anticipate, and trump, legislation.

It is generally agreed that any succeeding president, or a veto-proof majority of Congress, can overturn the declaration of a monument, just as a veto-proof majority of Congress can enact riders over presidential objection. No declaration of a monument, however, has ever been overturned by Congress or rescinded by a succeeding president, no doubt because doing so would be a very public antienvironmental action, whereas riders more often than not are removed before a bill is passed. The respective histories of these two strategies may well reaffirm the power of the executive, but, more significantly, they betoken the frustration policymakers often experience with the traditional legislative process and their persistent desire to achieve their goals, if necessary, by going around rather than through it. In President Clinton's case, there was the additional factor of his being in the final year of his second term. Republicans were not favorably disposed to his building an environmental legacy by this backdoor tactic, but as the President's term wound down, they seemed helpless to prevent it.

As relations between the two branches become more strained, they take increasing recourse to their stealth strategies. However serviceable the hallowed legislative process may be to formulate and effect our nation's policymaking during ordinary times, the very elements intentionally built into it to slow the enactment of proposed measures are those that can cause legislative gridlock when times get tough. When animosity between the two policymaking branches reaches significant proportions, it is the process, not the policies, that is often compromised.

CENTRAL IDEAS

The design of the legislative process is intended to be fair and open to all, so that the policies it enacts reflect broad public consensus, but its weaknesses have, in fact, been so exploited that, as practiced, it is dispositive toward certain interests and policy outcomes. Partly in response, new executive and legislative strategies—riders and monument designations—are increasingly used to circumvent the process.

Chapter 3

Environmental Regulation and the Reinvention of the EPA

> Administrative agencies of the regulatory kind are established to carry out the terms of the treaties that the legislators have negotiated and ratified. They are like armies of occupation left in the field to police the rule won by the victorious coalition.
>
> Earl Latham

> I'll let you write the substance on a statute and you let me write the procedures, and I'll screw you every time.
>
> U.S. Representative John Dingell

This chapter examines the broad and complex issue of environmental regulation, the aspect of the process that has the most impact on environmental policymaking and is, therefore, the most politically contentious. We first will look at the rules of the game—the formal procedures by which regulations are drawn up—since these procedures are dispositive toward certain kinds of policy outcomes. Then we will focus on the regulatory arena itself and the spectacle of the Environmental Protection Agency's (EPA's) turbulent history. As the principal institutional embodiment of environmental protection, the EPA has been the focal point of the myriad social, political, and economic forces that are exerted whenever any substantive environmental issue has been open for public consideration.

The Politics of the Regulatory Process

In the popular mind, and even in the minds of many who are actively involved in environmental policymaking, "rules and regulations" are synonymous with "laws"; the terms are routinely used interchangeably to refer to any environmental requirement. Yet rules and regulations, although they have the force and effect of law, are very different from laws, and their differences are significant. Understanding what rules and regulations are, who develops them, and the process by which they are adopted is crucial to understanding environmental politics and to effective participation in the policy arena.

What are the differences between regulations and laws, and why is it important to distinguish between them? Their differences are significant because virtually all of the major battles on environmental issues are fought over regulations, not laws. As noted in Chapter 2, most environmental statutes are, in effect, declarations of broad policy objectives: the air ought to be clean and free of contaminants that threaten human health; the waters of the nation ought to be at least "fishable and swimmable"; those who spill toxic chemicals on lands or into waters ought to be responsible for cleaning them up—with their money, not ours; development must be controlled because overdevelopment can increase the likelihood of droughts and floods and compromise the habitats of birds and other wildlife; the survival of certain species is tenuous, and it is important to do what is necessary to ensure that they do not become extinct; and workers must be aware of the dangers to which they are subject from unknown and complex chemical substances that they routinely handle in the course of their work. Because these are lofty and laudable objectives, they are not, by and large, controversial. Who, after all, would quarrel with any of them—at least publicly? This is why even the most far-reaching and ambitious of such laws—the *Clean Air Act* and the Superfund are two notable examples—were passed with healthy, bipartisan majorities.

This is not to say that the enactment of environmental statutes is a friction-free, amicable process. As we saw in Chapter 2, it is not. There is often much debate over whether the solution to one or another problem is properly the responsibility of government or whether the marketplace should be allowed to resolve problems in its own way. Even when the appropriateness of government action is conceded, there is often disagreement about whether the objectives in question ought to be pursued at the national or the state level. Disagreement, especially with respect to land use issues, also surrounds whether government involvement tramples on higher, sometimes constitutional rights, e.g., the right of landowners to use their property in any way they desire. Finally, even when all these issues are resolved, special interest groups vie for control of the agenda, clash over how a problem should generally be addressed, and work to get themselves excluded from the purview of whatever is being proposed. Except when statutes themselves have built-in "reauthorization" provisions (e.g., the Clean Air Act), or when a program's funding source expires (e.g., Superfund), or when a significant problem with the programs' implementation develops, laws seldom are revisited and the battle moves to the regulatory arena. Even the Reagan administration, widely regarded as the most antienvironmental in recent history and which announced as Reagan took office its goal to rein in environmentalism, concentrated on "regulatory reform" rather than statutory change. Indeed, the 1980 Republican National Party Platform acknowledged that "virtually all major environmental legislation reflected a bipartisan concern over the need to maintain a clean and healthful environment" even as it "declared war on government overregulation." Thus, the efforts of the Reagan administration to change statutes were few, modest, and tentative and were confined largely to "restricting public access to data" in the major pesticide control law and to relaxing standards and compliance deadlines in the Clean Air Act. Its assault on regulations, however, was direct and comprehensive.

What Are Regulations, and Who Makes Them?

Broadly defined, *regulations* are the specifically prescribed ways of carrying out the policy objectives of the laws. Legislators cannot, after all, be expected to know exactly what must be done to realize these policy objectives. They want to keep landfills from leaching, for example, but do not have the expertise to prescribe how they should be engineered to isolate the wastes from human contact. They want to ensure that nuclear power plants do not release radioactivity, such as occurred at Chernobyl and Three Mile Island, but they obviously do not know what risk-elimination systems are necessary or what training programs are appropriate for employees. The federal Clean Air Act goal of air relatively free of toxic pollutants has no opponents, but the control of contaminants that are discharged continuously into the air from hundreds of thousands of industrial and commercial activities and millions of cars involves perhaps thousands of scientific and technical determinations: standards for each statutorily identified contaminant, a protocol for automobile inspection and maintenance programs, a formula for reformulated gasoline, and the identification of specific technologies to minimize pollution at its source. Consideration must also be given to balancing costs and practicality against efficiency and effectiveness. All of these technical aspects are, of course, beyond the expertise of those who make the laws.

Much the same is true in the area of land use. The values of wetlands, and the need to preserve them, have also been embodied in law. How to define wetlands and what criteria to adopt for their designation, however, must necessarily be the business of experts. If the overarching question with regard to air is "how clean is clean," the question with respect to wetlands is "how wet is wet?" Do lands have to be veritable swamps to qualify, or do they simply have to support hydrophytic vegetation? How wide do buffer areas around them need to be to protect the resource? For that matter, how extensive an area is required to preserve the habitats of threatened or endangered species?

To realize their policy objectives, then, legislators need to turn to engineers, scientists, land use planners, lawyers, economists, and specialists in a variety of other disciplines to set the protocols, standards, methodologies, equipment specifications, tolerances, dosages, and time frames necessary to ensure that what they want to happen does happen. These are the "regulators," the personnel residing in the administrative agencies who must deal with the kinds of thorny questions enumerated above. The answers they provide are loaded with implications for a whole range of interests, but before addressing that issue, we must look at who these individuals are and the process by which regulations are developed.

We have seen that laws are enacted by legislators—officials elected every two, four, or six years and who are therefore directly accountable to those they represent at regular intervals. They thus have to be ever mindful of their constituents' interests and concerns. Regulations, on the other hand, are developed principally by administrative personnel who usually hold permanent full-time positions, often, though not invariably, protected by "civil service," the government counterpart to tenure in academia. To that extent, the line staff are personally insulated from external pressure. These staffers are, appropriately, selected for their technical expertise in a given area and are publicly invisible and anonymous. They develop their regulations collectively, in settings that are almost never covered by the press, and do not attach their names to their products. Unlike laws, regulations have no "sponsors."

It should be noted, however, that these regulatory personnel do work under the immediate supervision of administrators who have to sign off on draft regulations before they are offered for public comment. These supervising administrators are most often political appointees, and it is their job to reconcile individual regulations with the broad goals and policies of the specific agency as well as with the demands of their legislative sponsors, the President, and even the public. Their role is to maintain a larger perspective, if for no other reason than to ensure the viability and credibility of the agency's decisions. The tensions between core staff and administrators, as might be expected, are not infrequently strained.

Further, regulators are in an organizationally anomalous position. They are administratively located in the executive branch of government and work under the supervision of a cabinet secretary, administrator, or commissioner, but they are obliged to carry out the mandates of the legislative branch of government. This has not been a particular problem for most of our history, during which the two branches have typically been under the control of the same political party at any given time. However, in recent years, the public has become increasingly distrustful of politicians and has thus increasingly chosen to split tickets—electing, say, a president or governor of one party and a Congress or legislature of the other—to guard against feared abuses. In such situations, agency personnel are subjected to dual pressures and, in some cases, from partisan appointees in the agency itself.

The Regulation Adoption Process

Finally, let's look at the process whereby regulations are adopted. It ought to be noted at the outset that the statutes themselves often delegate rule-making authority to one or more of the relevant administrative agencies. Just as with the assignment of legislative bills to standing reference committees (see Chapter 2), legislators may well pick "friendly" agencies to draw up the rules. This is an additional measure of control that sponsors have over the implementation of their legislative proposals. The rules themselves are adopted in accordance with a process set forth not in the Constitution, as for laws, but in an umbrella statute known as the Administrative Procedure Act.

The process generally works like this: The delegated agency first reviews the statute to get a clear picture of its policy objectives. A steering committee decides what is in the purview of its responsibilities—how wide a net to cast, what expertise is required, what decisions have to be made, and what may need to be researched. It then creates a Working Group of staffers with the appropriate range of expertise. Members of the Working Group talk to people in and outside the regulated communities to get a feel for the potential practical problems and the costs of their ultimate prescriptions, collect relevant data, and begin work on the proposal. Because the members of the working group represent a broad spectrum of perspectives and knowledge bases, their interaction is itself a political event of which little publicly is known, but it often has significant impact on the final product, as we shall see in our discussion of the EPA later in this chapter. Their work results in a "formal proposal" to be submitted to the public for comment. In some cases, pre-proposals are developed and presented at hearings prior to rule-making. At the federal level, before proposed regulations can be submitted for public comment, however, they must be reviewed for fiscal impact and approved by the Office of Management and Budget, a requirement imposed by two executive orders. This process is extremely technical and complex and, in the case of major legislation, takes months, occasionally even years, to complete. The resulting draft is then submitted at public hearings at which anyone can testify and offer objections, suggested changes, compliance cost estimates, practical problems with implementation, and the like.

The regulatory agencies are legally required to submit their proposed regulations to this public review process. They are under no legal obligation, however, to change as much as a comma in response to anything brought up at these hearings, although they often have to prepare written responses to issues raised by the public. They can simply say that a given suggestion would not advance the purposes of the act, or is impractical, or is costly, and leave it at that. When the public hearing process on proposed regulations is complete, the agency formally adopts and promulgates the proposed regulations, which have the force and effect of law. At the federal level, rule-makers even have the authority to make substantive changes or additions without soliciting public comment, a privilege not generally accorded at the state level. Thus, negotiation of rules can and does take place internally, but it is in no way necessary to negotiate or compromise with interest groups after the hearing process has been completed.

Why Regulations Are Political

With a clear conception of what regulations are, of the administrative position and role of those who develop them, and of the regulatory adoption process itself, it is now possible to see why regulations are at the center of environmental politics.

First, it should be apparent that regulations, for all the science that goes into them, are inevitably policy documents. The establishment of any particular standard, tolerance, or concentration of an air pollutant, for example, may seem value neutral, but no matter what level is set, some populations will be affected more than others: children, the aged, asthmatics, or the chemically sensitive, to name but a few. Similarly, the extent of buffer areas to protect wetlands will have a different impact on commercial builders than on residential builders. Regulations concerning the level of cleanup of hazardous discharges from existing sites or buildings required as a precondition for construction or renovation is a major factor in any decision about urban redevelopment, as is the linchpin of the so-called "brownfields–greenfields" debate, i.e., whether a new facility should be sited on a previously developed site and thus promote urban redevelopment (brownfields) or be located on previously undeveloped land (greenfields). This is why the public hearing process on proposed regulations is at least as contentious as the legislative process. All manner of interest groups take the opportunity to voice their views. In the case of issues with far-reaching consequences, public hearings are held in several geographical areas to afford maximum opportunity for input.

The policymaking potential increases substantially in the case of state regulations to implement federal laws. As discussed in Chapter 6, the federal government delegates implementation of some of its major enactments, most notably in the areas of clean air and water, to the states. It does this principally because states have vastly different population densities, industrial profiles, geography, topography, and the like, and are in the best position to determine the least economically and socially disruptive means for them to reach the federal goals and targets. But by shifting its regulatory burden to the states, the federal government also escapes the political wrangling necessarily involved. As a result, the states have considerable latitude to apportion burdens among different interests, and incur a good deal more political pressure as a consequence. As we shall see later in this chapter, the EPA has recently gone even further in sharing rule-making responsibility with other government agencies and lower levels of government.

The process of rule adoption, then, is almost inevitably a lawmaking endeavor, even though most regulators would disdain such a notion and have no conscious inclination to serve as unelected legislators. Nevertheless, the nature of their work frequently makes them vulnerable to the charge of "making laws" without due authority, which puts them in an uneasy political position.

Next, let's look at legislators and the regulatory politics they provoke. In our examination of the lawmaking process in Chapter 2, we noted that it is characteristically a balancing act. Legislators start with a proposed solution to a public problem. To get their proposal passed, however, they must reconcile a variety of contending forces affected in various ways by that proposal. Compromise and trade-offs with their fellow legislators are the norm. Also, because they are under the constant watch of their constituents and function in a public forum with substantial media attention, they want to be—or at least want to *appear* to be—effective and reasonable as well as public-spirited. This often results in a passed bill with a complex and sometimes contradictory mix of provisions to accommodate or placate as many of those interests as possible. Their goal is to get the bill passed, and they often do whatever is necessary to accomplish that end. In their single-minded pursuit of enactment, they often shift inherent problems and inconsistencies to the regulators, who are ultimately responsible for making the bill work.

Worse, legislators like to get credit for "doing good," even if their socially conscious proposals will have an adverse effect on powerful interest groups. Although not often talked about, those inside the regulatory community can recount innumerable instances in which a legislator won public plaudits for sponsoring a praiseworthy bill and then, before the ink was dry on the president's or governor's signature, began pressuring the regulatory agency to "go easy" in its implementation in the service of an affected interest or constituent. Regulatory personnel, then, are unwittingly burdened with the "impurities" of the legislative process and often find themselves trapped between the political pressures from sponsors and their own integrity. Even if legislators do not actively coerce the regulatory agency into moderating its rules, they blame "bureaucrats run amok" when constituents come to them with costs and burdens flowing from the legislation. Regulators are often scapegoats for the downside of environmental statutes and even for the obvious and intended consequences.

Finally, let's look at the interest groups whose activities would be directly affected by the law. Many of them, of course, have weighed in during the legislative process, particularly those with formal representation and a substantial presence in the halls of Congress or state legislatures. Others find it unwise to press their opposition too fervently during the legislative process, particularly if, as is often the case, the bill purports to address a compelling public need. Such interest groups often wait until the rule-making phase, when they can pursue their interests more effectively—and clandestinely— at the regulatory table. The regulatory process affords the full spectrum of interests the opportunity to recover what may have been compromised, or conveniently left vague, during the legislative process or to cushion or undermine the impact of a given law on their activities. They may even use the regulatory process to achieve a competitive advantage in the marketplace by working to structure the regulations so that they hamper competitors more than themselves. Further, they can supplement their formal testimony with specific pleading to the supervisory administrators, who have the opportunity to incorporate provisions to accommodate them before approving a proposed draft. All this takes place in a setting that is, for all practical purposes, non-public, where they will suffer no bad publicity for appearing to challenge popular goals in the service of their self-interest. As a practical matter, the media are absent from, and largely unaware of, public hearings on regulatory matters. The process is too technical, protracted, and uneventful to be newsworthy.

A good example of interest group lobbying is found in the regulatory atmosphere surrounding state legislation proposed in the wake of the Bhopal catastrophe discussed in Chapter 2. The legislation granted the state Department of Environmental Protection extraordinary powers in the service of preventing a recurrence of any such tragic event. The state chapter of the national association representing the chemical industry was in no position to exercise its muscle to prevent its passage. The citizens of the state with the second largest concentration of chemical plants in the nation were understandably frightened by any possibility of a similar accident near them, and to oppose preventive measures would be, from a public relations standpoint, unthinkable. So they allowed the legislation to pass without any opposition, even though the powers it gave to the state to oversee chemical plant operations were unprecedented.

But when the state's Department of Environmental Protection began to craft regulations to flesh out the legislation, the chemical lobby participated actively. In "educating" the department about how chemicals were handled in sophisticated manufacturing plants, they were able to shape the regulations that plant operators had to follow henceforth. They were even consulted about which chemicals should be covered because the chemical released at Bhopal, methyl isocyanate, was by no means the only one that could have such an acute lethal effect on those exposed. Further, they were able to encourage acceptance of protocols that had the effect of giving the larger chemical companies, disproportionately represented by the association, competitive advantage over so-called mom-and-pop operations.

On the federal level, a similar strategy was employed by DuPont, the world's largest producer of chlorofluorocarbons (CFCs), the chemical compound associated with destruction of the ozone layer. When an ozone hole confirmed the adverse effects of CFCs on the earth's atmosphere, DuPont argued for, rather than against, a global policy to restrict them. No doubt perceiving that continued resistance would be fruitless and self-defeating, DuPont judged that limiting future production on a strict schedule would, in the short run, raise prices and maximize profits, especially since a unilateral ban by the U.S. would disadvantage them relative to European producers. More important, in the long run it would give them a head start in marketing a substitute, on which they had been working for years. Again, intra-industry competitive advantage guided the regulatory posture of DuPont.

When interest groups fail in their efforts to eliminate the regulatory provisions they find most repugnant, they frequently seek redress in the courts, arguing that their particular situation was not contemplated by the legislature in enacting the statute and that the regulations are therefore *ultra vires* and violative of "legislative intent." This charge, the most common raised in legal challenges to rules, essentially affirms that the regulators, whom they disparagingly refer to as "bureaucrats," are really acting as legislators and making laws rather than rules. In many cases regulators do go well beyond the reasonable parameters of a statute, but, as we have seen, the process of rule adoption is a policymaking activity. The courts must decide when regulators have crossed a critical line. Even when interest groups lose their cases in court, however, they have at least delayed implementation of the rules during the course of the case. That is itself a victory, because cases and subsequent appeals can take years.

In sum, interest groups often participate more actively and aggressively in the regulatory arena than on the floors and committee rooms of legislative bodies. The relative obscurity of the regulatory process allows parties adversely affected by popular legislation to pursue their self-interest without incurring public scorn, and the highly technical and sluggish pace of the regulatory process discourages media and public attention. The substantive expertise of many of these same players in highly specialized areas, however, also makes them valuable, sometimes necessary, partners in the rule development process.

Further, the administrative personnel who develop regulations are easier targets than duly elected legislators. Although they may seem to be insulated from political pressure by reason of their job security and their ultimate power over the final form of regulations, they are politically vulnerable in other ways, especially to the wishes of legislators. First, the legislature holds the power of the purse. Legislatures ultimately must approve the budgets of administrative agencies, and it takes little more than a threatened budget cut to get a "recalcitrant" agency to be "more responsive." The most pronounced exercise of this power took place during the Reagan administration, when the budget of the EPA, until that time on a steadily upward rise in response to ever-increasing responsibilities, was sharply reduced. Similarly, the 104th Congress, personified by House Speaker Newt Gingrich, stalled one of President Clinton's major proposals—to raise EPA to cabinet status, where it would have had more budgetary muscle.

That same Congress, in a little-publicized measure, The Small Business Regulatory Enforcement Fairness Act, granted Congress "legislative oversight powers." Specifically, a provision in the act called Congressional Review of Agency Rulemaking gave Congress 60 days to review major new regulations as a precondition to their taking effect and facilitated the passage of resolutions to disapprove those that it did not think satisfied "legislative intent" or that cost disproportionately more than the benefits it promised. Although such legislative authority has been called into question both on Constitutional grounds as violating the separation of powers doctrine and on the practical consideration that it would just give lobbyists another bite of the apple, President Clinton signed it in March 1996. Lawmakers also have sought to limit the discretion of regulatory bodies by including provisions in the legislation circumscribing their options in setting standards or establishing criteria or imposing fines and penalties for violations.

More recently, as discussed in detail in Chapter 2, Congress is responding to what they regard as a runaway EPA by including provisions as riders to unrelated legislation, many of which are regulations in all but name. In so doing, they can bypass the regulatory process entirely and provide low-profile relief to the regulated community.

Finally, it should be noted that with respect to regulation, in some ways legislators themselves are an interest group. They are concerned that rules adopted pursuant to their statutes do, in fact, implement the policy objectives of those statutes. That is their public obligation. But they also have a stake in minimizing the costs and burdens such rules may impose on their constituent interest groups. And so, like other interest groups, they add their voices and their influence to the many that routinely descend on environmental regulators.

In the mid-1980s a refinement of the rule-making process just described came to the fore: negotiated rule-making, or "regneg" for short. Regulatory negotiation applies the principles of alternative dispute resolution to environmental rule-making. Briefly, negotiated rule-making replaces the existing inherently adversarial process with one that allows interest groups to participate in the process before, rather than after, regulations are drafted. Doing so reduces the "thrust and parry" that characterize the traditional procedure, by allowing groups to put their concerns on the table early, rather than their hardened positions later in the process. Experience has shown that permitting affected parties to sit down with regulators at the initiation of rule-making minimizes costs by allowing parties to combine research on agreed-on issues, improves compliance, promotes public acceptance, and in many cases obviates the need for adversely affected parties to seek judicial review. Notwithstanding these advantages and its successful employment at the federal level in several environmental cases, the fact that negotiated rule-making has not become more widely applied probably indicates that political give-and-take is more deeply embedded in the regulatory process than anyone cares to admit. As we shall see, however, the Clinton administration made a specific effort to institutionalize regulatory negotiation in its "reinvention" protocol.

The Environmental Protection Agency and Institutional Politics

On March 16, 1995, President Bill Clinton and Vice President Al Gore issued a detailed report called *Reinventing Environmental Regulation* as part of their National Performance Review. In it they enumerated "10 Principles for Reinventing Environmental Regulation" by which the EPA would be guided in carrying out its mission in the future:

1. Protecting public health and the environment are important national goals, and individuals, businesses, and government must take responsibility for the impact of their actions.
2. Regulation must be designed to achieve environmental goals in a manner that minimizes costs to individuals, businesses, and other levels of government.
3. Environmental regulations must be performance-based, providing maximum flexibility in the means of achieving environmental goals but requiring accountability for the results.
4. Preventing pollution, not just controlling or cleaning it up, is preferred.

5. Market incentives should be used to achieve environmental goals whenever appropriate.
6. Environmental regulation should be based on the best science and economics, subject to expert and public scrutiny, and grounded in values Americans share.
7. Government regulations must be understandable to those who are affected by them.
8. Decision making should be collaborative, not adversarial, and decision makers must inform and involve those who must live with the decisions.
9. Federal, state, tribal, and local governments must work as partners to achieve common environmental goals, with nonfederal partners taking the lead when appropriate.
10. No citizen should be subjected to unjust or disproportionate environmental impacts.

On the surface, these principles appear reasonable, even innocuous. But against the backdrop of the EPA's history, they represent an almost point-by-point response to the succession of attacks leveled against it during the preceding decade by interests ranging from conservative and libertarian think tanks to environmental justice groups, as well as by Congress and the White House. Collectively, they represent almost a complete metamorphosis of the EPA's character, functions, and method of operation. Let's look at that history, because it reveals much about the political struggles inherent in regulatory policymaking in general, and about the complex and turbulent politics to which the EPA has been subject during the quarter century since its establishment.

Early History of the Environmental Protection Agency

The EPA was formally established by an executive order submitted by President Richard Nixon to Congress on July 9, 1970. The reorganization plan that created the EPA was the result of several yearlong struggles. The first was between presidential aspirant Edmund Muskie of Maine, then Chairman of the Senate Public Works Committee, and President Nixon, both of whom sought to enhance their election chances by corralling the newly emerging environmental constituency. There were also debates within the executive branch among a number of cabinet secretaries and agency heads whose powers would be diminished or compromised by one or another of the proposed reorganization plans being considered by the president. Finally, various congressional leaders were uneasy about the effect a shakeup in executive structure would have on their committees' jurisdictions.

The ultimate compromise involved the transfer to the newly minted EPA of a potpourri of programs, offices, and functions from other departments, a pair of councils, and a commission. With that mixed bag of responsibilities—for, principally, water quality, pesticide study and control, solid waste management, air pollution, and radiological health—came a somewhat eclectic constituency.

It should also be noted that in amalgamating existing agencies, the EPA absorbed personnel as well as an assortment of missions, working alliances, and perspectives. During previous years, staffers of those agencies had, of course, built up working relationships with the regulated entities under their respective jurisdictions and, inevitably, certain sympathies for their concerns and needs, which they took with them to their new "home."

The executive order's preamble clearly established pollution control as the new agency's central responsibility, to be carried out by more coordinated research, standard-setting, and enforcement, while offering to industry consistency of standards and to states the promise of financial and technical assistance. Environmental interests hoped finally to have a strong, consolidated advocate for their causes in the executive branch; the president and other federal officials saw in their creation of the EPA a means to manage environmental protection in a way that accommodated the needs of business and industry. Thus began a debate over just how to raise the child, so to speak—how much independence to give it, how to meld the disparate influences on it, how to control its spending, and how to punish it for "misbehavior."

The early years, during which the agency sought to gain its footing, found it courting environmental interests. The EPA's first administrator, William Ruckelshaus, focused initially on strict enforcement rather than on organizational structure or technical and scientific resources. This predisposition gave rise, however, to the first of what would be a succession of attempts to rein in the EPA; the Office of Management and Budget (OMB) established as preconditions to regulatory adoptions "Quality of Life Reviews," a euphemism for a requirement that the EPA consider economic development and fiscal concerns in its regulatory process. Ruckelshaus's challenges to this authority moved the EPA away from White House control and closer to Congress, a direction continued by his successor, Russell Train. Douglas Costle, who was appointed EPA administrator by President Jimmy Carter, felt the need to better relations with the White House, and the president wanted to begin to put substance onto his claim for better government. Accordingly, President Carter, by executive order, established the Regulatory Analysis Review Group, a clear descendent of the Quality of Life Review process, which would mandate that all federal agencies conduct an economic impact analysis of each proposed regulation that had a projected cost of $100 million or more a year.

It is important to note at this point that attempts to rein in the EPA were in large part a reflection of two countervailing factors. The first was its growing responsibility and authority. The 1960s and 1970s, and even much of the 1980s were, from an environmentalist's perspective, halcyon days. Fueled by a series of environmental disasters—Times Beach, the oil spill off Santa Barbara, Love Canal, Three Mile Island—public fears and anxieties impelled Congress to pass a score of major pieces of legislation, each requiring strict and far-reaching regulatory implementation. But with the EPA's growing power came a regulatory burden that was simply enormous and unmanageable. Not only did these laws require extremely complicated and sophisticated technological and scientific determinations, but they also imposed strict timetables for their development and promulgation. The EPA fell months, even years, behind schedule, which disturbed Congress, the environmental community, and in some cases the business community. Therefore, the pressure on the agency to perform increased as its budget, level of personnel, and flexibility decreased. Yet its power and authority over an ever-widening range of problems made it perhaps the most feared of federal agencies. It became an easy scapegoat for the economic malaise that marked the Carter years and, thus, the centerpiece of Ronald Reagan's election platform.

The Reagan Assault and Its Aftermath

Ronald Reagan's comfortable victory over President Carter was generally viewed as an affirmation of the public's desire for smaller, less burdensome, and less expensive government, and in large measure it was. But the public was not yet ready to reduce its commitment to environmental protection, as the administration was soon to learn.

President Reagan's environmental strategy was crafted by conservative think tanks—principally the Heritage Foundation—whose ideology is market oriented. There were several components to that strategy, but perhaps the most significant and enduring was implemented pursuant to Executive Order 12291, which created a Task Force on Regulatory Relief and required proposed regulations to be submitted to the OMB for final approval before they were presented for public comment. The OMB was also given authority to initiate agency action. Reagan's predecessors had incrementally expanded the role of the OMB, but Reagan made it a major obstacle before any proposed environmental regulation could go public. It was in the deliberations of the OMB that another component of the strategy—cost-benefit analysis and risk-benefit analysis, long sought by business interests—were integrated significantly into public policymaking. These were hurdles not easily cleared by EPA, since costs are readily calculable but benefits often are not. Also, as has been frequently noted, the OMB was beyond the reach not only of administrative agencies, but also of the courts and Congress, all of whom were loathe to interfere in the work of the president. The tug of war between Congress and the administration over the EPA clearly swung in favor of the White House. Consistent with the probusiness bias of the president, meetings of the OMB as well as the EPA were opened to the regulated community and substantially closed to the range of public interest groups that had commonly been allowed to voice their views in previous administrations. Many EPA officials of had previously worked in the private sector and were thus quite accommodating to them as policies were discussed.

A second component of the antienvironmental strategy was devolution. Reagan sought to limit the power of the EPA by transferring power and authority to state and local governments. Federalism, as it is commonly called, involved delegating implementation of federal laws to state governments wherever possible and to developing generic laws that states could implement in their own way. Reagan also encouraged and promoted privatization, whereby traditional governmental activities and functions would be contracted for with the private sector.

Finally, President Reagan significantly reduced the budget of the EPA, despite the fact that, as noted, it was striving mightily to try to keep up with its delegated responsibilities. Its sharply reduced funding necessarily limited the EPA's research capacities, even as it was paradoxically required to justify its rules on sound science, turning it more into an enforcement agency. Some of these actions even rattled the administrator of the EPA, Anne Gorsuch and were among the factors leading to her premature departure from the agency.

President Reagan did not accomplish the emasculation of the EPA that he seemed to contemplate, but he did leave it wounded—short on funding, low in employee morale, and subject to administrative encroachment by other agencies with different agendas and different constituencies. Even though, under the second stint of William Ruckelshaus, it regained some of the respectability it had lost during the scandals and excesses of Reagan's first term, the EPA, during the administration of George Bush, was still reeling.

The late 1980s and early 1990s were a period of fiscal austerity for the nation, and the case was often made, if not always persuasively, that environmental regulation was a drain on jobs in particular and the profits of business in general. And so it was virtually without public outcry that Vice President Dan Quayle could chair a Council on Competitiveness comprising a group of business interests—the Secretaries of Treasury and Commerce, the Attorney General, the Director of the Budget, the Chairman of the Council of Economic Advisers, and the Chief of Staff—which, acting behind closed doors and with no public record of proceedings, heard the complaints and concerns of private-sector executives. During its tenure, the council unilaterally overturned scores of regulations deemed by the vice president as excessive, burdensome, and unnecessary. The essential point is that the Council fundamentally ignored the established legal procedure for adopting regulations by simply vetoing regulations that it felt imposed costs on business disproportionate to their anticipated benefits. An outraged Congress sought to contain the Council by reducing its budget, but it nevertheless continued until it was disbanded by President Bill Clinton. Thus, during the administrations of Reagan and Bush, government became extraordinarily sympathetic to business interests, a shift facilitated by a struggling economy that was pinned to environmental excesses.

In an effort to survive, the EPA, during the latter years of the Reagan administration, began work on new initiatives, not the least of which was one to prioritize its workload. In 1987, the EPA's Office of Policy Analysis published *Unfinished Business—A Comparative Assessment of Environmental Priorities*, a compilation of task force reports establishing agency priorities based on risk. This represented the first major effort at coordinating the various programs in the EPA, at once raising science to a higher level and providing some rationale for allocating limited money and resources in publicly defensible ways. It also involved, in a significant way, local units of government.

That work, however, failed to forestall the second major attack on the EPA, that launched by the 104th Congress, representative of a Republican sweep in the 1994 elections. There were at least two major prongs to that attack. The first was, at least superficially, benign. Congress commissioned the National Academy of Public Administration to determine whether the EPA was "allocating resources to the most pressing environmental concerns" and to determine how and by whom priorities were set. The charge was implicitly stimulated by an alleged loss of trust in the effectiveness of the EPA by the public, even in the face of continued public support for its mission. *Setting Priorities, Getting Results,* as the resulting 1995 report was called, was, as one might expect from its authors, a management study. Its conclusions, also not surprisingly, were strikingly similar in direction to that the Reagan and Bush administrations wanted to pursue: (1) a more defined mission, but the flexibility to carry it out; (2) devolution of responsibility and decision-making authority to the states and localities; (3) flexibility and accountability to the private sector and to local governments; (4) improvement of its management operations by establishing specific goals and strategies, using comparative risk analyses to inform priorities, and setting and tracking benchmarks of progress; (5) expanded use of risk analysis and cost-benefit analyses in making decisions; and (6) a breakdown of its media orientation in favor of a more integrated structure.

The Contract With America: Echoes of Reagan

The other prong of the attack was more stealthy. The Republican House candidates had campaigned collectively for election on a preannounced program of legislative initiatives that it called "Contract With America." One of the major components of that program was a bill called the Job Creation and Wage Enhancement Act. Although that particular bill, as it was described in the Contract, did not even mention the word *environment*, it contained three measures that would, if enacted, significantly undermine environmentalism: (1) a requirement that all federally mandated requirements on the states be funded by the federal government, the so-called "federal mandate, federal pay" provision; (2) a requirement that owners of land on which development is restricted by environmental laws be fully compensated for the loss of their economic value; and, tellingly, (3) a requirement that virtually all environmental regulations be based on sound science, a requirement that was to be implemented through submittal by the EPA of the scientific premises, studies, and conclusions on which the regulation was based to "the public," including the regulated community, for complete review as a precondition to their promulgation. The obstacle that the Republican constituency sought to create was not only substantive but practical as well: it would take enormous resources to collect, systematize, and distribute such voluminous data, and it would be prohibitively time-consuming to review the comments from the outside world. Actually, the House of Representatives, the driving force behind this provision in particular and the "Contract With America" in general, passed some version of each of the three provisions, but only the federal mandate federal pay was passed by the Senate and ultimately signed by the president. Neither of the other two were passed by the Senate, presumably because its president, Bob Dole, was by then seriously considering a run for the U.S. presidency and couldn't brook a nationwide debate about such volatile and far-reaching issues.

It would have been reasonable to assume that the new stresses on the EPA that the Reagan and Bush administrations exerted—to be less autocratic and more cooperative, consumer friendly, and market-oriented—would abate with the election of Bill Clinton as president and long-time environmentalist Al Gore as vice president in 1992. In a curious way, though, they did not. Indeed, they were sanctified and institutionalized more firmly in the government protocol than ever before. This was attributable to several factors.

First, the now almost decade-old critique of government as bureaucratic, inefficient, indifferent, and costly, initiated by President Reagan was, by and large, adopted by the general public. Average citizens did not see any inconsistency between this characterization of government and their confidence that this same government, in the name of the EPA, could protect them from toxic waste, polluted air and water, and urban sprawl. Each citizen seemed to bring to consciousness his or her own nightmarish experiences with government, a notion supported by the popularity of two books that commanded nationwide attention.

The Changing Politics of the Clinton Administration

Reinventing Government by David Osborne and Ted Gaebler, a carefully crafted prescription for better governance that was published in 1992, implicitly capitalized on this underlying discontent. Osborne and Gaebler presented a measured mission and protocol for a new kind of government, one that "steered" instead of "rowed" and was competitive with the private sector, mission driven rather than rule oriented, customer friendly, preventive rather than curative, decentralized, and market-oriented. The authors seemed to show disdain for political labeling, but their recommendations were uncannily close to, if not congruent with, those advocated by the anti-EPA forces of the previous several years. It also provided what was to become a significant term in the regulatory reform vocabulary, *reinvention*.

Another book on the same theme was more unsparing and critical. *The Death of Common Sense* by Philip Howard, published in 1994, was an openly antiregulatory tract that detailed a litany of the consequences of expensive, overcomplicated, misguided, and ineffective government rules, many of which, though by no means all, were environmental. Its thrust was purely negative and sarcastic—providing countless laugh lines for luncheon speakers at business gatherings nationwide—but it clearly struck a chord. After all, how often does a book on government sit atop the *New York Times* best seller list for the better part of two years as this one did? It, too, provided an important term in the regulatory reform vocabulary. The notion that the EPA's priorities and protocols often violated "common sense," the mantra until now of the antienvironmental community, was soon to be sanctified and institutionalized by the new Democratic administration.

While the enemies of the EPA were eroding its credibility from the right, a long-standing issue—the disproportionate number of unwanted facilities whose siting was approved by the EPA in low-income and/or minority neighborhoods—came to the fore in the early 1990s. The EPA had been defending itself from charges of racism for more than 20 years, but the environmental justice movement finally coalesced with the People of Color

Leadership Summit in October 1991 and the "Principles of Environmental Justice" document that it produced. The event made environmental justice a potential political land mine for the EPA, impelling it to modify its previous self-righteous stand as a dispassionate scientific agency to one that would henceforth be sensitive to equity concerns. But it was not until President Clinton issued Executive Order 12898 in 1994, directing all federal agencies to incorporate environmental justice concerns into their decision-making process, that such interest was given formal recognition.

Pressure not only from partisan forces and environmental justice advocates, but also from the popular media, including a steady stream of articles in newspapers and magazines literally charting the disparity between what the experts thought were the most serious threats to public health and the environment and the EPA's agenda, impelled President Clinton and Vice President Gore to embrace the principles of *Reinventing Government* and recast the EPA. The previously mentioned March 1995 report announced a new era in environmental regulation. Henceforth, the EPA was to seek "common sense" (the phrase is repeated countless times in the report) solutions to environmental problems, keep an eye to cost-effectiveness, and, perhaps most important of all, share decision making with those affected by its actions. The EPA would henceforth focus its regulations on performance and market factors; base priorities on sound science; build stronger partnerships with the private sector as well as state and local governments; reduce paperwork; improve accountability, compliance, and enforcement; and, in general, become more user friendly.

Specifically, the report presaged the initiation of nearly two dozen new experimental programs, one important piece of which was called, not incidentally, the Common Sense Initiative (CSI). In the words of the EPA Administrator Carol Browner, the CSI is an "effort to make good on what business and environmentalists have been telling us for two decades—that we must look at whole facilities, whole industries, and their overall impact on the environment. We must do a better job of cleaning up the environment and do it cheaper." Accordingly, the CSI pilot program convened teams of representatives from all levels of government, environmental interests, labor, industry, and environmental justice groups to study existing regulations in six industries. The teams were charged with developing regulatory protocols that would promote pollution prevention rather than "end of the pipe" controls, were flexible, would focus on whole industries rather than on individual pollutants, and were consensus based. Thus, the media-based, pollutant-focused "one-size-fits-all" regulator was to begin its transformation into a collaborative, results-oriented facilitator.

In succeeding years, the CSI program has waned, but a companion program, Project XL (Excellence and Leadership), which offered the regulated community a straight swap of regulatory flexibility for improved environmental performance, has become the centerpiece of EPA reform. The program invites facility operators, industry sectors, communities, and even government agencies regulated by the EPA to submit proposals for exceeding environmental standards using strategies that would otherwise violate existing regulations. The primary criteria for approval of an XL project—better results, cost and paperwork reduction, promotion of multimedia or pollution prevention effect, and stakeholder support—are established by the EPA, but the central purpose of the effort is to decentralize regulation. The EPA's role is limited to establishing the program's criteria, working with the states to negotiate the projects, monitoring the results, and extracting lessons that may be applied in other regulatory contexts. Although environmental interests are concerned with a potential lack of accountability, less public participation, and a circumvention of judicial review, they are, for the moment, holding their fire until definitive results are available.

And so the 30-year history of the EPA has seen it transformed from an authoritarian, rigidly bureaucratic, centralized power to a more flexible, pro-business, cost-conscious, power-sharing facilitator. That transformation was embodied in microcosm in the "10 Principles for Reinventing Environmental Regulation." It was spurred by battles between successive Congresses and presidents and the constituencies each branch of government wanted to use the EPA to cultivate, by public attitudes toward government in general and its impact on the economy in particular, by consistent pressures from the regulated community exerted directly and indirectly through legislative and regulatory agents, by the insistent and compelling claims of economically disadvantaged and minority ethnic populations, by the various needs and situations of individual states and municipalities, and by forces within the agency itself. The history of the EPA is, finally, a mirror of the complex mix of political forces at work virtually since the inception of modern environmentalism.

Still, antienvironmental forces refuse to trust this transformation, or at least trust its permanence. No fewer than a half dozen bills were introduced in the 106th Congress that would embody in statutory law the institutional changes that the EPA has made in recent years. These include the Regulatory Improvement Act of 1999, which would require federal agencies to perform risk assessment and cost-benefit analyses for major pieces of legislation; the Regulatory Right-to-Know Act of 1999, which would require the OMB to estimate the costs and benefits of each federal regulation; the Mandates Information Act of 1999, which would expand the unfunded mandates law of the 104th Congress by requiring a statement disclosing the effects of proposed health, safety, and environmental legislation on the regulated community; the Federalism Act of 1999, which would make it more difficult for federal laws and regulations to preempt those of states and localities; and the Congressional Accountability for Regulatory Information Act of 1999/2000, which would require, among other things, the General Accounting Office (GAO) to review economically significant federal rules and to conduct further cost-benefit analyses of each rule, alternatives to the rule, and the effects of lower levels of government. Although the Reagan administration may well have failed to effectively undermine the federal regulatory apparatus, its agenda lives on more than a dozen years later.

CENTRAL IDEAS

For a variety of reasons, most of the significant environmental policy is made at the regulatory level rather than at the legislative level. The EPA, the agency principally responsible for developing, implementing, and enforcing environmental law, in response to new political pressures, has transformed itself from an authoritarian, centralized bureaucracy into a flexible, collaborative facilitator.

Chapter 4

The Media Business and Environmental Politics

I'd sit in on meetings about the Clean Air Act and experts and scientists would tell us in detail why taking certain steps to reduce air pollution slightly wasn't worth the cost in dollars and jobs. . . . Then the media would do stories that were mostly emotional and had little to do with the facts I had been hearing. And people would start talking about changing their vote.

Ernie Schultz

There is a tension between the scientific culture of caution and reticence and the media's penchant for drama, dread, and debate that keeps the show lively and the audience tuning in.

Stephen H. Schneider

Journalists are almost invariably disturbed, if not scandalized, by any allegation that they are political. By and large, members of the news media see themselves as disinterested purveyors of information, and their professional code regards bias as heresy. And yet, in a profound and complex way, the media have been, and continue to be, political. In no other area is this truer than for the environmental arena. How the media's role as political actor evolved, why it continues, and what its implications are for the making of environmental policy is the subject of this chapter.

To call the media political is not to suggest that print journalists, news anchors, and television reporters—to say nothing of editors and producers—do not genuinely aspire to the estimable goal of dispassionate, objective communication. With few exceptions, they do. But the reality is that the larger context of news in our society, the way it is produced, and the forces that shape it all pose formidable obstacles to the realization of their best intentions. And environmental issues, for a variety of reasons, make those obstacles almost insurmountable. This circumstance has far-reaching and profound implications for environmental policymaking, especially considering that the public that drives the environmental agenda is generally unaware of the inevitable limitations of the environmental reporting on which it depends and which colors its views of what is happening "out there."

Environmental journalism and the environmental movement virtually grew up together. We should not forget that Rachel Carson's *Silent Spring*, generally regarded as the clarion call to the modern environmental movement, was first published in a magazine, *The New Yorker*, in 1961, and the shocked and dismayed reaction of its readers made the formal appearance of *Silent Spring* in book form a year later a literary event of epic proportions. The story of *Silent Spring* was not only the launching pad of environmental journalism, but, in many ways, a paradigm of the genre: the dramatic disclosure of the harmful, potentially lethal effects of a common practice imposed on an unsuspecting public by a powerful, indifferent industry. With only the details changed, it was a story that would be told countless times in newspapers and magazines and by the broadcast media for the next four decades.

At the core of the media's mission are two obligations: to inform their readers and viewers of events and circumstances that are important if not essential for them to know about; and to earn a profit, because press outlets and broadcast networks are, after all, businesses and, as such, must maintain their economic viability. Although these two mandates are by no means mutually exclusive, there is an inherent tension between them, and editors, publishers, and producers have had difficulty seeking their reconciliation.

The reconciliation was not difficult to achieve during the 1970s and 1980s. The Love Canal and Times Beach chemical incidents, the Chernobyl and Three Mile Island nuclear malfunctions, the Santa Barbara and Valdez oil spills, and the toxic release at a Bhopal, India industrial plant were all events of national, and in one case international, significance that caused the death, illness, property damage, and relocation of thousands of people. They were blockbuster stories, and because most of America's population lives relatively near one or another factory, toxic or solid waste facility, nuclear power plant, potential oil drilling site, or other site that could pose the same or similar threats, they raised fears on the part of the public for their own safety and the security of their neighborhoods, to say nothing of property values.

The media milked these events not only for their environmental implications but for their drama and dread as well and, in so doing, drove the legislative agenda while at the same time increasing their own circulation and viewership. *Time* magazine rang in the 1980s with a cover story on the pervasive and growing threat of toxic chemical wastes, which it called "The Poisoning of America" (September 22, 1980) and closed the decade by making "Endangered Earth" its 1989 "Planet of the Year" (January 2, 1989). *Time* was not alone in mining public anxiety over the environment. Even such stolid journals as *National Geographic* asked on the cover of its special double issue of December 1988, "Can Man Save This Fragile Earth?" *Scientific American* published its own special issue, "Managing Planet Earth" in September 1989, to be followed on April 30, 1990 with the *New Republic*'s "The State of the Earth" issue, the cover of which pictured a greenish black funnel cloud bearing down on a lonely, helpless man. And so the decade closed amid a barrage of apocalyptic soundings. It is not difficult to see why the principal catalyst for the halcyon days of environmental activism was the public's call for government action to protect them from the many ominous threats they were hearing so much about.

Fortunately, events like the disastrous ones of the 1970s and 1980s virtually disappeared in the 1990s, to be replaced by environmental issues that were more subtle, more technical, and more chronic than acute, though no less real and significant. Having raised public consciousness about environmental risks, the media were constrained to find in such matters—or, failing that, to generate—some of the same interest that the big events of the previous decades had commanded. Those events had "made news" by themselves. Now "making news" was left to writers, anchors, and production staffs.

Making Environmental News

The major disasters and foreboding circumstances earned the environment a place on the news agenda, alongside politics, finance, sports, entertainment, health, and consumer affairs. But absent disastrous attention-grabbers, the environment had to take its place in line and fight for space and time against competing subjects. This meant that it had to become "news" in its own right. For the environment, this was, and remains, an especially challenging task, because almost every characteristic of most environmental issues goes against the grain of what has come to be regarded as "hot copy." The way environmental reporters—and others interested in advancing the environmental agenda—have tried to meet this challenge has not only altered the nature of environmental coverage, but, in so doing, has also drastically changed the politics of the environment. Let's look now at how this has played out, and then examine some of its consequences.

Although the public's concern for a clean and safe environment has never wavered, as poll after poll has confirmed, its interest in media coverage of environmental issues has waned, if editors and producers are to be believed. In their minds the environment has gone from nightmare to "snore," the word they use to characterize its newsworthiness. Why are most environmental subjects so antithetical to engaging journalism? The reasons range from the narrow limitations of the news genres themselves and the prevailing practices of news production, to the many demands they place on environmental reporters, to a professional ethic to which journalists are committed, to the pressures to shape stories coming from so many different sources. Let's examine these factors one at a time.

First, it is almost axiomatic that, in an age of "infotainment" and shortening attention spans, any story has to engage the audience on more than an intellectual level. This is especially true as a steadily increasing percentage of the population gets their news from television rather than newspapers. Television news has had to combine the traditional story line and relevance to the audience with strong visuals, which neither radio nor newspapers could do. Environmental events can satisfy these prerequisites—at least some of them can—and that is where the media as a political force has its origin.

Selecting News Stories

It might be best, then, to start with the criteria for what journalists call "newsworthiness." If stories are not deemed newsworthy, they don't get into the press or on television in the first place, no matter how important they may ultimately prove to be. Journalism textbooks are pretty much agreed on the basic ones. A principal criterion is *conflict*. Where there is no conflict, there is no story line, and a narrative flow is an important element of contemporary journalism—a way to hold the diminishing attention spans of the viewing or reading public, and increasingly important as "sound bite" journalism competes with the traditional forms. Indeed, the use of the word "story" to identify the genre is of relatively recent vintage. It was not so very long ago that what appeared under headlines in the newspaper were routinely called "articles," and their first paragraphs just as routinely answered the "five W's—who, what, where, when, and why." Today, they are almost always called "stories," and, accordingly, often begin anecdotally, with the introduction of particular people in a situation emblematic of the larger one that is the main subject of the piece. Such a change in treatment, as we shall see, plugs into a number of the criteria of newsworthiness, but its principal value is to take what the readers may regard as dry and academic and make it into a mini-drama that they may someday find themselves involved in.

There is no shortage of conflicts in the environmental area, which is a principal reason that it can hold its own against sports, finance, and politics, provided, however, that the issues are chosen carefully and pitched the right way. The conflict between the rights of private property and government, for example, may be introduced by homeowners who are shocked to learn that they cannot rebuild a deck on their house because it encroaches on a wetland that has been identified and mapped since the house was built. The battle between development interests and environmental activists may well take the form of marchers protesting the siting of a major new subdivision in an area that is vulnerable to flooding. Conflicts between competing industries, between different segments of the same industry, and even between feuding scientists can be explored, but they are deemed more engaging to the reader if the history of how they developed, of who is in the middle of the fray, and of how it is likely to turn out is reported as well. But many big issues—global warming, the loss of biodiversity, overpopulation, ozone depletion—cannot be easily presented as conflicts and have no identifiable victims, nor are they readily cast as narratives, since their progression is, for all intents and purposes, imperceptible and their "beginnings" and "endings" are highly speculative. That is why they do not get the copy that their significance merits. They are, rather, circumstances that must be revisited periodically as new information becomes available or as conditions change. There is simply no place in the newsroom milieu for tracking continuing developments.

Another criterion of newsworthiness is *novelty*. Again, the public is almost invariably attracted to the unusual and the bizarre and less to conditions or circumstances to which they have become inured. This accounts, at least in part, for the ubiquity of articles on genetically modified foods, for example, as opposed to the diminishing number on air or water pollution. *Timeliness* and *proximity* are other criteria. Whatever the substance of any particular concern, the public can be captured more easily by dangers or alleged risks that are present or impending, e.g., those posed by cell phones or microwave towers, than by those that are far off in distance or time. The public doesn't have a long horizon—witness their relative equanimity in the face of a growing body of scientific evidence confirming predictions of a cataclysmic global warming over the next century. Much the same can be said of threats posed to their own communities or lifestyles. Local news is clearly of more interest to them than threats far from home that have no immediate impact. This may well account in large part for the "domestication" of the policy agenda spoken of earlier in Chapter 2.

Public interest in media stories is also enhanced by *prominence*, by the involvement of celebrities. Actor Robert Redford's considerable allure has given credence to any number of environmental causes, and the infamous Alar brouhaha of a decade ago was much more widely publicized because of the active participation of Meryl Streep than it otherwise would have been. Actor Woody Harrelson's efforts on behalf of old-growth redwood forests and fellow actor Ted Danson's activities to raise public awareness about ocean water quality have brought public attention to otherwise silent causes. On the same theme, in the wake of the blockbuster movie *Titanic*,

Leonardo DiCaprio, its star, made headlines as an ordinary citizen concerned about global warming, but his heavily advertised, eagerly anticipated interview with President Clinton was all but aborted because, in securing the interview, he upstaged the professional journalists, who apparently could not generate comparable publicity for the issue.

More broadly, a whole cadre of entertainers have identified themselves with a wide range of environmental movements and causes, but whether their motives are public-spirited or self-promotional is in dispute. Formal organizational initiatives seek to capitalize on this phenomenon as well. The Environmental Media Association identifies itself as the premier organization of celebrities supporting the environment, whose mission was, in part, to get appropriate messages into films, television, and commercials. Its awards program specifically recognized environmental messages in movies such as *Dances With Wolves*, *A River Runs Through It*, and *Free Willy*. Finally, the Earth 911 Promotions Group, whose hotline provides environmental information to citizens, gained much wider recognition when Julia Louis-Dreyfus, a star of the popular television comedy sitcom *Seinfeld*, promoted it. Her efforts allegedly secured more than 2,000 new members for the Environmental Defense Fund.

These criteria for newsworthiness—the dramatic, the new, the bizarre, the timely, the local, the glitzy—may or may not represent what is most consequential, but are, nevertheless, the factors that determine what people read and hear about most frequently. Environmental issues suffer disproportionately by their application. The broad scope and reach of many significant environmental issues are often short-circuited, their issues oversimplified, and their effects trivialized by these criteria.

All this is not to suggest that social and environmental impacts are not themselves sufficient to "make news." Environmental happenings and conditions that affect large numbers of people in serious ways do find their way onto the airwaves, but even such phenomena need to be carried by some element of immediate, tangible consequence. For example, in recent years, global warming has most often made it into the news when it was arguably implicated in a drought or a flood, or when climatological statistics suggested that most of the hottest years in history were in the 1990s. Similarly, depletion of the ozone layer gets fresh attention whenever an increasing incidence of skin cancer is reported. In short, the public seems to want their environmental stories to have some immediate relevance or be of some practical consequence—"news you can use," so to speak.

The "Construction" of News Stories

What makes news is also conditioned by the form in which stories are cast and the practices of newsroom production. Stories in the print media are presented in column inches, and on television in time slots—neatly delineated packages. While this factor constrains the presentation of any but the simplest subjects, it is especially fatal to most environmental problems, which are more often chronic than acute and which commonly require scientific and/or economic background to be understood. Environmental stories are complicated, sometimes abstract and technical. The complex, inherently interdisciplinary nature of environmental issues does not easily accommodate itself to the space and time limitations of media presentation, nor are audiences prepared to give them the concentration they require, at least not when they are reading the newspaper or watching television.

Further, because of the necessary limitations on space or time in which they are presented, news stories are inevitably self-contained, discrete entities. This requires that they be "constructed" in the newsroom, that they be "framed," and, as they say in the business, given an "angle." Events that can be scheduled and preformulated have a leg up on those that cannot. All these considerations obviously work to the disadvantage of environmental problems, because the most important are usually amorphous, have multiple rather than singular sources, and are also rife with implications not only for the immediate situation at hand, but also for a host of large social issues, e.g., the proper role of technology, the relative demands of the economy and the environment, or the competing claims of social good and individual freedom.

Many of the same factors that make environmental subjects unattractive to media coverage impose enormous burdens on the environmental journalists themselves. First, environmental law is a huge and complicated labyrinth of statutes and regulations that is beyond the grasp of the general public and all but the relatively few professionals involved in it. Explaining science to a lay public and making concepts such as risk-benefit analysis accessible to the everyday reader or viewer are daunting tasks indeed.

Second, the environment's interdisciplinary nature places unusual demands on reporters, so much so that there is a continuing debate over whether environmental journalists ought to be specialists, have their own beat, or at least have some training or special course work to prepare them for the myriad issues they will face. Such training would, presumably, enable them to understand and evaluate the information they get from the experts they interview in researching the stories. That element of their job is itself a challenge. There are, of course, the usual institutional sources to whom reporters can, and usually do, go to first—government officials with the relevant responsibilities as well as the appropriate politicians. It is essential, however, that they also seek out representatives of relevant private-sector interests and of environmental organizations to fill out their stories. Scientists also must be consulted when an issue warrants it. Each of them almost invariably brings his or her own interpretation of the problem or event, and it is the job of the journalist to tease out the hidden agendas from the facts. To ease the difficulty, journalists often build up a list of trusted sources from whom they regularly seek out background and information. But there rarely is enough time to examine a problem or event in detail and consult people with the relevant range of expertise.

The whole issue of sources introduces the factor most responsible for putting the media at the center of environmental politics—ironic, because they brought it on themselves. As noted earlier, the journalism profession maintains, as an article of faith, the mandate of objectivity, or at least fairness, but pursuing that ideal has its consequences. In their effort to achieve balance, reporters are obliged to seek out and publicize the views of people on both sides of the issue at hand. (I use the word "both" advisedly, because the notion of a multiple number of sides in public disputes is anathema to Americans. As a society we tend to view things from a dual perspective—liberal or conservative, guilty or innocent, right or wrong, safe or dangerous.) In an effort to get "the whole story," journalists endeavor to be fair to all involved, which obliges them to report fully, and without editorial comment, what "both sides" say and think, however responsible or irresponsible—or isolated—a particular spokesperson may be. Thus, stories about environmental risk tend to represent opposing views rather than truths, leaving the determinations about the latter to the audience. Reporting the clash of views not only protects the journalist's claim to objectivity, but provides an element of conflict as well. Thus, even on issues where the weight of scientific opinion seems to be disproportionately on one side, conflicting versions of the truth are afforded virtually equal coverage. Again, global warming is a good example. While most scientists regard the phenomenon as sufficiently likely to justify a systematic and deliberative public policy response, skeptics are routinely represented and quoted in virtually every story on it, thus giving the minority view equal coverage and weight.

The interest groups know all this, of course. Environmental organizations, whatever their honest appraisal of priorities, provide the media with a steady stream of scare stories about possible or prospective catastrophes, and thus keep the environmental pot boiling, as well as build their membership and contribution levels. The media, for all the reasons noted above, welcome them. Antienvironmental interests, even as they bewail the media's compulsive attraction to crisis, exploit the media for their own purposes. They know very well that reporters will present their opposing positions, however well founded, to balance the coverage. Thus, our newspapers and television and computer screens serve as battlegrounds for dueling positions and ideologies on a wide range of environmental matters.

Superficially, there is nothing wrong with the media's serving as a forum for debates on environmental issues and happenings. Indeed, that is what they genuinely think they should be doing. But it should be obvious from the foregoing discussion that the environmental subjects they choose to write about and the way they treat them engages their audiences in a profoundly emotional way. And the public concern generated by such stories thus finds its way to their elected representatives, who take media stories quite seriously, however solid their foundation. The historical correlation between the environmental issues that have been accorded a high profile and the body of environmental law enacted by Congress is unmistakable.

The forces shaping the selection and handling of environmental stories have other substantial political effects as well. It is a journalistic truism that "bad news drives out good news." For whatever psychological reason, people don't think that the communication of progress is "news." Thus, while new threats to public health or the environment quickly find their way into the public awareness, the real successes of programs designed to attack them do not. Thus, the important but unspectacular improvements in air and water quality over the past several decades, reductions in solid and hazardous waste generation and improvements in their management, the achievement of significantly greater energy efficiencies across the board, and the de facto ban on nuclear power plant construction, to mention only a few examples, remain unacknowledged and unreported by the press and thus generally unknown to the citizenry. This not only distorts the public's perception of the state of the environment, but also implicitly undermines their perception of the effectiveness of government's ability to solve environmental problems and manage risks.

Even more serious, media coverage that exaggerates minor risks in pursuit of the sensational and the bizarre diverts public attention and resources from real ones. Policymakers and regulators are continually frustrated by their having to allocate limited money and personnel in ways that they know are counter to good sense and good science in response to public perceptions and pressures, while more serious problems go begging. For example, the relatively minor health and safety risks posed by high-profile Superfund sites have been systematically addressed at both the federal and state levels for more than a decade, and at great cost, whereas the more substantial but clearly more low-key threat of indoor air pollution, especially of radon infiltration of homes, has largely been ignored. Time and time again over the past decade polls have graphically delineated the sharp disparities between what the public thinks is dangerous and what the experts think dangerous. This disconnect between perception and reality can be attributed only to the media stories that created these perceptions in the first place. The consequences of the skewed priorities manifest in public policy are profound.

One more circumstance regarding media coverage of the environment ought to be at least mentioned here, though it will be treated more extensively in Chapter 9. I am speaking of the consolidation of media companies and their acquisition by multinational corporations. This development has imposed additional commercial pressures on them. Producers and editors now have to satisfy not only their historical audiences and sponsors, but stockholders as well. Again, this is a fact of life that most news areas have to deal with, but the threat to environmental reporting is perhaps more serious than to the reporting of other matters. As globalization takes hold, an increasing number of environmental issues take on international implications, and the transnational corporations have deep interests in how they are treated. The effect on U.S. environmental standards of international agreements such as NAFTA and GATT, on its economy by treaties such as those signed at Rio de Janeiro and Kyoto, and on major U.S. biotechnology corporations such as Monsanto by pacts such as that signed at Cartegena governing trade in genetically modified foods all affect the manner in which many issues can be covered. It is still another obstacle that environmental reporters have to steer around in covering the environment.

Everything in the current culture of journalism, then, works against the responsible, comprehensive, and thoughtful presentation of environmental problems, circumstances, and developments. The prevailing criteria of newsworthiness, the unique demands of newsroom practices, and increasing pressures from new interests such as shareholders all contribute to an editorial policy framework that is antithetical to the character of environmental issues. Few environmental issues are inherently dramatic or newsworthy, and almost none comfortably fit into prescribed time slots or column inches. They are complicated, scientific or technical, and generally protracted. Their sources are multiple, their effects are long term, their underpinnings are interdisciplinary, and their consequences frequently expose the financial interests of the very outlets that give them currency.

It is the synergistic effects of this set of circumstances and mandates that have inexorably drawn environmental journalists into the vortex of politics. Despite their best efforts to stay above the fray, the environmental media have been understandably vulnerable to the persistent charge that they are "closet greenies." The factors described in this chapter predispose journalists to write about threatening events—some real, some exaggerated, some frankly nonexistent—in an effort to attract audiences and satisfy commercial demands. In doing so, they have unwittingly served as the handmaidens of environmental interests. The media have thus constituted a conduit for environmental groups to Congress: by publicizing these matters in the dramatic terms that they have, they have generated a level of public concern that policymakers have not been able to ignore. The media, then, have become a principal catalyst for an activist environmentalist agenda over the past four decades, even as they have not recorded as faithfully as they could have the environmental gains achieved by that activism.

However strongly media coverage of the environment has influenced Congress, it has incurred the displeasure, if not the wrath, of two other interest groups. Scientists, who share some of the same obligations to the public as journalists but work in vastly different ways, often believe that the media misrepresent much of their work. According to a recent comprehensive survey, almost two thirds of scientists think the media exaggerate the risks associated with various activities. Four of five find the media more interested in "instant answers and short-term results," and three of four believe that sensationalism is a higher goal than is scientific truth. These are no doubt the reasons that almost half of the respondents prefer to avoid the news media altogether because they are "suspicious of their motives." Again, one can see the hand of "newsworthiness" in shaping these attitudes, but the split with the scientific community it occasions has significant political implications, particularly when matters get to the policy stage.

The Media and Antienvironmentalism

Of even greater political import, the disaffection with the environmental media that scientists feel has been exploited persistently by the journalists' principal detractors—a loose consortium of antienvironmental interests comprising conservative economists and political scientists such as the late Julian Simon and Aaron Wildavsky, "Wise Use" types such as Ron Arnold and Alan Gottlieb, libertarian ideologues such as Michael Fumento, conservative think tank publicists such as Ronald Bailey, antigovernment popular scientists such as Dixie Lee Ray, and talk show hosts such as Rush Limbaugh. For a variety of reasons, these people and the larger philosophies or constituencies they represent are making careers out of opposing the government's historic and deep involvement in the environmental area. One has even revived a journalism career by launching a glitzy and well-publicized attack on journalistic treatment of environmental issues in a national TV special in 1994. In "Are We Scaring Ourselves to Death?" ABC correspondent John Stossel characterized himself as a reformed sinner, a former consumer affairs reporter who used to exploit exaggerated risk stories for their human interest, who has since come to see the error of his ways. His engaging production attributed the distorted priorities of our environmental polices to the triumph of fear over science and economics. These antienvironmental forces have sought to discredit prevailing environmental polices by discrediting the environmental media, painting them as "sky-is-falling" alarmists who seek out catastrophe for its marketability and who force needless and costly government action by foisting apocalyptic prophesies of doom on a fearful public. That they can, and frequently do, enlist the testimony of scientists in support of this indictment only makes their case stronger.

But even as they chide the media for serving as shills for environmental groups, these antienvironmentalists are not above exploiting the power of the press themselves. For they also know that the media can be relied on to balance reports about dangers posed by new chemicals or technologies, or about the promise of proposed new laws, with their own opposing perspectives. After all, balance, not accuracy, is the overriding value, particularly for the regulated media such as radio and television. And so, in the best traditions of both journalism and politics, these contrarian voices are respectfully accorded "equal time."

Thus, the environmental media have, ironically, become an interest group themselves, playing an active role in the political arena. Unlike most of the other interest groups, they do not come to political table with any particular policy objectives. Rather, the communications climate in which they operate, the practices and mandates of their profession, and—perhaps most important of all—the commercial pressures to which they must always be responsive, clearly give them an interest in environmental politics and make them influential forces in the shaping of environmental policy, for better and for worse.

CENTRAL IDEAS

The ideal to which the media aspire is objective, nonadvocacy reporting. Yet, by virtue of the criteria of newsworthiness, the practices of newsroom production, and the professional commitment to "balance," the media become, unwittingly, interest groups and agenda setters.

Chapter 5

Uncertain Science—
Uncertain Politics

> Environment is one-tenth science and nine-tenths politics.
>
> Anonymous British Delegate
> U.N. Conference on Human Environment

> The public has become used to conflicting opinion.
> . . . Many have come to feel that for every Ph.D., there is an
> equal and opposite Ph.D.
>
> Tim Hammonds

Scientists, like journalists, see themselves as standing apart from the world of politics. The scientific enterprise has historically been imbued with an aura of objective authority, and the laboratory setting, where scientists give free rein to their curiosity, is at the opposite end of the world from the legislative chamber, where society's immediate problems and needs direct its course of inquiry. The systematic pursuit of ultimate truths, the freedom from coercion, the self-imposed isolation from the events, circumstances, and pressures of the day—these represent the context in which scientists ideally like to work. But it is an ideal that is difficult, if not impossible, to realize, particularly at this time in history, and especially in the areas of environment and ecology.

For a variety of reasons, many not dissimilar from those affecting the media discussed in Chapter 4, science and its practitioners have, wittingly and unwittingly, become deeply entrenched in the environmental politics of the day. Though the research and experimentation that constitute the substance of science continue at ever-increasing levels of sophistication, the people, institutions, and processes that determine what is to be researched, who will perform the research, and to what purposes it will be put are political matters. It is over these issues that environmental scientists are increasingly being drawn into the fray, both as subjects and as participants. It may, in fact, be argued that not since the Darrow-Bryan debates over creationism in the mid-1920s has science been so publicly on trial, its very integrity, as well as its capacity to provide the necessary underpinnings to environmental policy, being challenged almost daily. There may be no more telling index to the plight of environmental science than the fact that the 1990s saw the publication of two books with the identical title of *Science Under Siege*, but written from, and reflecting, opposite sides of the ideological spectrum that their respective authors, Michael Fumento and Todd Wilkinson, occupy. Science is, in fact, under siege and has been for more than a decade. Why and how this has happened, what battles have been fought and to what effect, and just how environmental policy has been and will continue to be affected by the struggle are the subjects of this chapter.

At the core of the matter is the simple but significant fact that every environmental problem has, at its foundation, a scientific reality. It therefore seems axiomatic that science must play a prominent, if not pivotal, role in formulating its solution. This is why every interest group from the most ardent environmental activists to those who would have free market forces dictate environmental behavior all support the notion that environmental policy should be guided by "sound science," though they differ radically on how they define that term. Whatever their differences, "sound science" is the buzzword of the times. Everyone agrees that, given the growing complexity and impact of potential problems, and the social and economic costs that may be necessary to address them, the need to enlist science in making public policy decisions has never been greater. The unfortunate fact, however, is that science and public policymaking are fundamentally incompatible. A disjunction events between the practice of scientists and the needs, demands, and expectations of the other major players in the political game—lawmakers and regulators, the courts, the media, and the population at large. Science and scientists thus find themselves fighting a political war on several fronts simultaneously.

Science Is Analogue; Public Policy, Digital

Let's look first at the relationship between scientists and lawmakers, who work in very different ways and have very different missions. The scientific method is a measured, incremental, and systematic process of identifying a problem, collecting data, developing a hypothesis, establishing a tentative conclusion, testing it, and submitting the results and the methodology to others working in the same area. This practice is called *peer review*. Hypotheses must not only be verifiable, but falsifiable, i.e., capable of being tested so that predicted outcomes will occur if they are true, or fail to occur if they are not. This process takes time and depends on self-imposed doubt. Thus all scientific conclusions are tentative, the product of a collective meeting of minds, with other scientists observing and studying the same or related matters. While definitive truth is, of course, an ultimate objective, it is rarely arrived at. Scientists generally require agreement among 95% of the relevant experts to establish cause and effect, a level that is seldom reached. More important, scientists not only are comfortable with uncertainty, they literally build it into their thinking.

It doesn't take much imagination to see how scientists and the scientific method frustrate the lawmaker and the policymaking process. They represent totally different cultures. Lawmakers are faced with problems for which the public demands prompt solutions and cannot wait for definitive data, whereas science is patient and tentative. Science develops gradually and changes, but once public policy is set, it is hard to modify. Differences of opinion among scientists are simply part of their everyday world, but lawmakers see such conflicts both as complicating their decisions and as but engendering public distrust, thus making anything they do suspect. Lawmakers understandably want a degree of certainty that scientists cannot provide. They find that scientists have a tendency to overqualify conclusions to the point of uselessness or, more commonly, to avoid making recommendations entirely, preferring to remain separate from politically charged issues. On the other hand, if they are looking for science to sanction what is their disposition to do politically, they have no difficulty finding respectable scientists to support them, whatever their position may be. Finally, politicians are put off by the inability of most scientists to simplify complex issues and so make them understandable to laypeople.

Scientists, for their part, are reticent to participate in the political process. They feel that they are in fact most often brought into the policymaking process to sanction already decided policies, rather than to help formulate them. They find lawmakers not only scientifically illiterate, but, more important, intolerant of uncertainty and unappreciative of the concept of probability. They see them as wanting black-and-white answers, often without apprising scientists of the context of the problem being addressed.

Beyond these issues, however, are fundamental differences between policy and science themselves. Public policymaking is priority setting and involves considerably more than obedience to the bedrock scientific facts underlying problems. Science can, at its best, inform policy by telling us the relative probability of an environmental threat or estimating the likelihood of a particular consequence of that threat. Science cannot, however, set political priorities; it cannot tell us how we should allocate our social and financial resources to meet these threats or whether they are more or less deserving of attention than a host of problems in other areas. And science certainly cannot predict with any certitude the relative success of alternative courses of action to address them.

The uneasy relationship between scientists and lawmakers, then, and the disjunction between their respective missions and how they go about pursuing them make science's role in policymaking tenuous at best. The obstacles are all the more intractable in the areas of the environment and the earth's ecology because of the interaction of a staggering number of forces—natural and manmade—that synergistically impact the earth; because so many of the problems, potential and present, are global in scope; because many are nascent rather than manifest, and thus have delayed consequences; and because, in measuring their impact on human life, traditional trial-and-error experimentation is ethically unacceptable. Environmental scientists therefore have to develop their conclusions from computer models, statistical extrapolations, educated projections, and often just plain conjecture. For these reasons, their work in this area is inherently more technical and clouded with uncertainty and therefore more vulnerable to second-guessing.

Congress's Love-Hate Relationship with Science

It is against this backdrop that the political conflagration in which science finds itself today can best be understood. Because Congress is the locus of policymaking, the recent history of its complicated relationship with science is a useful place to begin. As noted earlier, the period before the Reagan administration represented a burgeoning of environmental activism during which literally scores of environmental laws were enacted, each requiring a complex body of regulations. The demands they imposed on the scientific capacity of the EPA and its sister agencies, academic institutions, and industry subcontractors was enormous. As Chapter 2 suggests, Congress, in its zeal to "do the right thing," but not really knowing the regulatory implications thereof, directed the EPA, under unreasonable deadlines, to implement its laws. There was simply no way that those regulations could enjoy the benefit of thorough and reliable scientific analysis and data. The result was a potpourri of different standards—technical feasibility, public health without consideration of cost, and best available technology, to name a few. In an era when environmentalism was so popular among the public, few were willing to challenge the tentative, shaky scientific foundations on which so many of these regulations were based.

But as the costs of compliance grew with increasingly stringent standards, that scientific foundation became the target of the regulated community. It was, not surprisingly, the Reagan administration that made the first move. By advocating the primacy of risk assessment and cost-benefit analyses in environmental rulemaking, Reagan was challenging the EPA to justify the science on which its regulations were based, and dragging science itself into a policymaking role. At the same time that he was publicly raising the stakes for what he regarded as good science, however, he reduced the EPA's scientific resources by cutting its budget. While he met with only modest success, at best, Reagan clearly challenged the soundness of the science upon which environmental policies were developed and implemented. Largely as a result of his actions, EPA scrambled to refine its agenda and sharpen its scientific focus, to establish defensible priorities, and to begin to measure success by empirical environmental indicators.

The partisan war on science was dramatically intensified by the 104th Congress under the speakership of Newt Gingrich. Picking up the mantle of Reagan, this Congress, in its "Contract With America" and in conformity with the Republican Party platform, again made risk assessment and cost-benefit analysis centerpieces of its regulatory reform effort. Specifically, the platform required regulatory agencies to "require peer reviewed risk assessments based on sound science," but even as it trumpeted the need for basing regulations on "sound science," Congress announced a series of initiatives to sharply reduce the scientific resources of the executive branch. The efforts were buoyed by a 1994 poll released by the Advancement of Sound Science Coalition, a national watchdog group of academic, government, and business members, which found that almost two thirds of scientists believed that public confidence in scientific research had decreased over the preceding ten years and that more than four of five thought that policymakers used science only to achieve their preestablished objectives. It was the kind of support that the antiregulatory Congress needed to justify proposals to abolish the U.S. Geological Survey, the National Biological Service, and the U.S. Bureau of Mines, as well as to substantially reduce the budgets of other federal agencies, most notably the National Oceanic and Atmospheric Administration. Interior Secretary Bruce Babbitt likened these proposals to "book burning."

Although none of the proposals came to fruition, on September 30, 1995, Congress did abolish the nonpartisan Office of Technology Assessment (OTA), a 23-year-old executive branch agency whose mission was to assist Congress in dealing with the increasingly complex technical issues that affect society. Although the public justification for doing so was that OTA guidance was not timely enough to inform pending legislation and that the information they provided could be secured from other sources, the action seemed inconsistent with Congress's own stated principle. As Robert T. Watson, Associate Director for the Environment at the White House Office of Science and Technology Policy, said in an interview in the June 9, 1995, issue of *Chemical and Engineering News*: "The Hill is asking [the Houses of Congress] to look at comparative risk and cost-benefit analysis. . . . good ideas basically. . . . But comparative risk and cost-benefit analysis require good knowledge, good science, and good technology. . . . In eliminating OTA, Congress wants to cut out the only independent, bipartisan scientific and technical group that provides effective advice for comparative risk." By this action, Congress had, at least on the surface, raised questions about its own commitment to "sound science."

Those questions became even more pointed when, on the heels of the demise of the OTA, the Subcommittee on Energy and Environment of the House of Representatives Committee on Science convened a series of three hearings on "Scientific Integrity and Public Trust: The Science Behind Federal Policies and Mandates." Subcommittee Chairman Dana Rohrabacher and Committee Chairman John Doolittle, in their introductory remarks, said that the purpose of the hearings was to examine the question of whether the nation was "getting objective science from our regulatory agencies," but it was transparent that the majority had as their agenda the undermining of the science that had, in fact, already gone into federal policy in the areas of stratospheric ozone, global climate change, and dioxin contamination, the subjects of the three hearings.

These hearings not only broadened the scope of the attack on environmental science from the narrow regulatory issues associated with cost-benefit analyses to more global ones, but set forth the central line of attack that Congress and its antienvironmental constituencies would level at the EPA for the remainder of the decade and beyond: that overzealous regulators were squandering public dollars on a politically driven agenda. As Representative George E. Brown, Jr., former chair of the Science Committee and ranking minority member on the committee, described in his report in response to the hearings: "They [the Subcommittee] implied that scientists themselves were part of a vast conspiracy with environmental regulators. The terms of the pact were that the scientists would exaggerate their certainty and consensus on environmental problems and the bureaucrats could use these statements, with help from their environmental activist allies, to push through ever more stringent regulations and ever greater funding for the researchers. Bureaucrats were funding science that justified their existence and scientists sold their integrity to the bureaucrats in exchange for steady funding."

As one of the House's few science experts, Congressman Brown felt compelled to issue a sharp rebuttal to the majority's claims. The rebuttal took the form of an extensive report titled *Environmental Science Under Siege: Fringe Science and the 104th Congress*. It attacked the explicit mistrust of government science displayed in what were advertised as fact-finding hearings. It also exposed what to Congressman Brown was perhaps its most curious complaint—that the views of scientists outside the mainstream of scientific thought were not allowed to influence policy decisions, implying "that scientific truth is more likely to be found at the fringes of science than at the center." Brown saw this as a "disturbing repudiation of the scientific process and peer review . . . inimical to a constructive role of science in policymaking." Further, he saw the hearings themselves as advancing the dangerous notion that a congressional committee could, in fact, serve as a science court, capable of determining scientific truth through testimony and questions and thus promoting the very politicization of science it was condemning. Finally, Brown's report took exception to what he regarded as the overall message of the hearings—that "sound science is empirical science," implying that statistical analysis and models are "speculation" and that sound science resides exclusively in empirical fact or observational data.

The hearings and the response they engendered have had little discernible effect on policymaking. But they do throw into bold relief the essential arguments of both sides of the debate over how science has been, and should be, used in the policymaking process. The congressional majority, representing not only their own ideological disposition but the many business and commercial constituencies opposed to environmental activism, have tried to exploit the inherent "weaknesses" of science as a policy tool—principally its inability to provide the certainty that lawmakers claim they need to impose financial and social burdens on the public. Absent this certainty, the underpinnings of those policies are, in their minds, reduced to mere hypotheses and speculation or, worse, the hyperbolic claims of fear-mongers who would capture government and commit it to their "save-the-world agenda." Environmentalists, on the other hand, argue that, when it comes to phenomena such as global climate change and stratospheric ozone depletion, or even the epidemiological consequences of exposure to air pollutants or pesticides, definitive conclusions are impossible, and to require certainty as a precondition to action is but a thinly veiled attempt to stifle regulatory activity altogether.

In what may well have been an attempt to save face, Speaker Gingrich commissioned the House Science Committee to review the country's science policy and specifically to write a sequel to the 1945 report, *The Endless Frontier* that had guided U.S. science policy for decades. In September 1998, the Committee released its report, *Unlocking Our Future: Toward a New National Science Policy*. Although its main recommendation is that the U.S. commit to stable and substantial funding for basic research, it does touch on the hot-button issue of peer review in specifically trying to fund "creative . . . speculative" studies that wouldn't be likely to be peer reviewed. Perhaps more significant, Congressman Vernon Ehlers, the principal author of the report, joined with 90 of his colleagues in cosponsoring a bill to establish a National Institute for the Environment under the National Science Foundation, dedicated to improving the scientific basis of environmental decision making. The Institute was not created legislatively, but it was brought under the umbrella of the National Science Foundation as the National Council on the Environment in February 2000, where it is currently pursuing its core mission to "improve the scientific basis for environmental decision making."

For all the homage paid to science in the legislative process, it is at the regulatory level—principally at the EPA—that science actually plays its most significant role. In fact, the credibility of rules and regulations depends on the public confidence that our laws are being administered in observance of the best and most rigorous scientific principles. But, again, that expectation is not, and cannot, be fully realized.

First, the EPA is not a scientific agency. Its principal charge is not to determine how best to attack environmental problems, or even which problems to attack, but to implement and enforce Congress's mandates. Though it does maintain scientific offices and advisory boards and committees, its statutory obligations take precedence over the generation or evaluation of new information. Accordingly, the EPA employs more attorneys than scientists, and its decisions are based as much on economic, political, or administrative considerations as purely scientific ones. Finally, because of the tight deadlines imposed by Congress to carry out the laws, its regulatory agenda cannot drive its research agenda, making it heavily dependent on the science developed outside, rather than inside, the agency—some from consultants, some from previously developed research, and some from either the environmental community or the regulated community itself. And the inability of science to provide definitive and precise answers to many regulatory questions has led both Congress and the EPA to seek political protection increasingly in prescriptive technology-based standards. All of this makes the EPA justifiably vulnerable to claims that its decisions rest on shaky scientific foundations.

The Courts' Struggle with Science

It is appropriate to briefly mention the challenge that environmental science poses to the courts. In most ways, the courts are like legislatures, i.e., they are faced with specific cases that need prompt resolution; they cannot wait indefinitely for definitive information. And like legislatures, their standard for decisions, except in criminal cases, is "preponderance of the evidence," closer to the 51% majority required to pass legislation than the 95% agreement among peers that scientists require to establish a scientific fact. Further, like legislators, judges and juries are faced inevitably with the prospect of sifting out the truth from a series of question-and-answer sessions with equally credentialed experts on opposite sides of the issue. Like their counterparts in the legislature, they are scientific laypeople and thus largely incapable of arbitrating technical disputes.

Courtrooms are no better suited than committee rooms to resolve scientific issues, but the Republican platform of 1996, in its section on "restoring justice to the courts," commits its party to "[e]liminate 'junk science' by opportunistic attorneys by requiring courts to verify that the science of those called as expert witnesses is reasonably acceptable within the scientific community. . . ." Although the pledge addressed the political objective of minimizing or eliminating huge monetary awards such as those enjoyed by plaintiffs in some recent high-profile cases, it is unrealistic to expect courts to do this, for the reasons noted. Proposals to establish specific courts to hear environmental cases and to provide their presiding justices with specialized training in relevant subject areas have not been adopted. The practical problems have seemed overwhelming, and the results far from assured. For both the legal system and the lawmaking process, an overriding concern is a social stability that derives from consistency of policy, which pure science cannot promise. As one critic put it, "science policy" is an oxymoron. And so the courts, though free of partisan pressures, also struggle to incorporate "sound science" into their decisions for most of the same reasons as their counterparts in the legislative and executive branches of government.

Science and the Media

The political role of science in public policymaking is by no means confined to the halls of government. The relationship between science and the media is as problematic as that between science and lawmakers. As noted in Chapter 4, scientists are commonly distrustful of journalists. Again, it's a question of incompatible missions. Because of the primacy of "newsworthiness" in the world of communications, what the media hope to get from scientists are breakthrough developments, pioneering findings, significant firsts, and bold predictions of significant consequences resulting from environmental conditions or threats. But the cautious, plodding, tempered pace of science seldom provides journalists with what they want, and scientists are generally not given to public pronouncements.

Because the public gets virtually all their scientific information as well as their environmental guidance from the media, and because scientists who do want to affect public policy most often speak to lawmakers through the media rather than directly, the split between them has serious political consequences. Indeed, the title of a 1997 study conducted by the First Amendment Center characterizes scientists and journalists as Worlds Apart, a circumstance that "threatens America's future." Despite the best intentions of people on all sides of the issues, science is having a hard time constructively informing national policies. The incongruity of missions is only one relevant factor, however. Again, like lawmakers, journalists generally do not have the training and experience to understand the technical language of science, and scientists are often reluctant to provide them with explanations for fear that the nature and significance of their work will be oversimplified and presented out of context. Further, editors and producers, if not journalists, see themselves not as educators but as chroniclers, and thus refuse to serve as publicists for research work. Consequently, an uneasy truce prevails between them, and the public is left to sort out what is relevant and meaningful to them.

Of course, there is no shortage of offers to help the public reach its own conclusions on scientific matters related to policy issues. It is no doubt unfair and unproductive to speak of "the media" as though they constituted a monolith. While major city newspapers, general circulation magazines, and national networks must appeal to broad, general audiences, "the media" include, as well, a wide range of publications across the ideological spectrum, and these have participated actively in environmental politics. The broad media outlets welcome them, since they engage in just the kind of conflict that makes for lively disputes.

One of the most politically active of such publications is, surprisingly, *Science*, the official organ of the American Association for the Advancement of Science. While its articles are invariably of high quality, peer reviewed, and policy-neutral, *Science* has not been hesitant in its editorials and in its letters column to give vent to a whole range of opinions on political issues. Its first agenda item, as might be expected, is to advance the role of science in public policymaking, and several of its columns in recent years suggested ways to do that. Individual columnists and any number of letters to the editor, however, have critiqued both ill-advised applications of science in public policy and the "flight from reason" that critics of science display.

Clearly to the right of *Science* are the house organs, newsletters, and research papers of a number of think tanks and foundations promoting free enterprise and unregulated markets—the Cato Institute, the Heritage Foundation, the Reason Foundation, and the American Enterprise Institute, to name but a few. They provide a platform for out-of-the mainstream scientists with estimable credentials and, usually, university affiliations, but whose work is nevertheless rarely peer reviewed. S. Fred Singer, president of the Science and Environmental Policy Project, is the most visible and influential of these since the death of Dixie Lee Ray, but a cadre of his colleagues, e.g., Robert Balling, Hugh Elsaesser, Richard Lindzen, and Patrick Michaels, constitute a group of prolific scientists who consistently challenge the proponents of environmentalism and whose work is consistently picked up in the editorial pages of big newspapers and on talk shows, such as that of Rush Limbaugh.

The environmental community also has its own organizational support. Two of the more well known of these are the Union of Concerned Scientists and the Physicians for Social Responsibility. Each of these groups has attracted a strong scientific membership, some among the most prominent in their fields. Yet their activist policy disposition is transparent and influences other professionals in the field as well as the general public.

It should be noted that each side has reinforced its policy positions—and rebutted those of its antagonists—in greater detail in a number of popular science books. Dixie Lee Ray's *Trashing the Planet* and *Environmental Overkill*, Ronald Bailey's anthology *The True State of the Planet*, Michael Fumento's *Science Under Siege*, John Baden's anthology *Environmental Gore*, and Ben Bolch and Harold Lyons' *Apocalypse Not* have gained the most publicity as antienvironment science tracts. Counterpointing them have been books such as Barry Commoner's *Making Peace With the Planet*, Paul Ehrlich's *Betrayal of Science and Reason*, and a spate of works published under the auspices of the Worldwatch Institute. And so science has been the target of the right and a weapon of the left in a continuing media war for the public mind.

Given the respected reputation of science in general among the populace, it is not science itself but its exploitation that is the issue between the two camps. Each accuses the other of misusing, even abusing, science to serve its own agenda. It is therefore not surprising that a term for science appropriated for political purposes— "junk science"—has gained currency in this climate. Coined by Peter Huber in the 1980s, it entered the public lexicon in the early 1990s, and has since become an epithet routinely hurled by both pro- and antienvironmental interests across the ideological divide of environmental activism. The Washington Legal Foundation, one of the think tanks whose mission is to promote free enterprise and less government, calls it "phony science concocted to further activist regulatory agendas and profitable litigation." The Union of Concerned Scientists, on the other hand, has defined it as "the 'data' and 'research' that some corporate interests and radio talk show hosts have been force-feeding America." There is even a popular website, junkscience.com, whose motto is "All the junk that's fit to debunk," an obvious spoof of the *New York Times*'s "All the news that's fit to print." The site is a collection and exposé, of what it regards as claims of phantom or exaggerated risks appearing in the popular media. Its creator and editor, Steven Milloy, an adjunct scholar with the Cato Institute, not only manages the site but writes editorials both for the site and for a variety of newspapers mocking the ungrounded fears they cause and the motivations of those who plant them. It has already spawned a companion site, NoMoreScares.com, "dedicated to following the misanthropic adventures of Fenton Comunications," the public relations firm retained to publicize such highly volatile issues as those surrounding Alar, silicon breast implants, bovine growth hormone, and endocrine disrupters. More significant, Milloy now has an established link with a national network. He writes a regular internet column for the Fox News website, foxnews.com/science/junkscience/index.sml. Junk science has apparently established a place in the media mainstream.

Obviously, the term "junk science" does not disparage science itself, but it does indicate how sensitive the competing sides in the environmental wars are to claims that science is on the side of their opponents. By disparaging the sponsors and motivations of the scientific findings ostensibly supporting their antagonists' proposed policy directions, the two sides hope to knock the props out from under each other's position and preempt, to the extent possible, any public support for action based on their scientific assertions.

Underpinning the charge that science and scientists are policy driven is that science has no constituency of its own, partly because of the reluctance of most scientists to have their credibility compromised by political involvement and partly because so much of their work and funding have historically come from the necessarily secretive defense sector. As defense as a national priority has receded and as environmental issues have concomitantly risen to the fore, however, scientists are looking to new sources of support. A substantial number of scientists work for government as employees or, by contract, for academic institutions with government grants, or by heavily regulated industries. In none of these capacities can their work be considered policy neutral. That a large percentage of the funding for science in general, and for environmental science in particular, comes directly or indirectly either from government sources or from those regulated by government is sufficient to give credence to the competing claims of antienvironmental interests that science is being put to activist purposes and of the environmental community that corporations and their hired scientists seek to roll back regulations.

Hence, scientists find themselves in a precarious position. On the one hand, they want to publicly establish themselves as experts in a relevant field of research and thus be in line for research grants and projects. On the other hand, prematurely going public with scientific conclusions may put them substantively at odds with institutional policies or industry positions that would disqualify them from those very same opportunities. Still, the paucity of sources of income from purely dispassionate research, as well as public pressure to come down from their ivory towers to help solve our problems, increasingly pressure scientists to communicate more through public vehicles. Dorothy Nelkin may have put the resolution best: "Scientists ventriloquate through the media to those who control their funds." Whatever the perils, increasing numbers of scientists are recognizing the importance, personally as well as professionally, of talking to the media and making themselves available to policymakers. Their political isolation is grudgingly coming to an end.

Public "Uncertainty"

It remains to look at the challenge science faces in enlisting the support of the general public, at whom all these appeals are directed. The uncertainty that is at the source of science's alienation from lawmakers and the media no doubt unnerves the public as well, who encounter it in their daily lives. The past several years have witnessed a see-sawing of scientific opinion—duly noted in the press—on such matters as the threats posed by radon, by dioxin, and by MTBE (the gasoline additive required to meet Clean Air Act standards), each of which has been downgraded and upgraded by turns as a genuine environmental threat. On an even more mundane level, there have been decade-long debates over the relative environmental consequences of paper and plastic grocery bags, of Styrofoam and paper beverage cups, and of cloth and disposable diapers. It is hardly surprising that a public that can't seem to get definitive answers to superficially simple scientific issues would question science's ability to understand global phenomena scores of years before their manifestation.

Of course, scientists are not that uncertain about radon, or dioxin, or MTBE and can surely help us with our everyday environmental choices. Their answers, though, which are usually relative to the medium and duration of exposure or the environmental problem to be alleviated, are not welcomed by a citizenry that is by and large scientifically illiterate and not receptive to less than black-and-white responses. This is one reason that politics continues to prevail over substance in the environmental wars. Another is science's ability to detect pollutants in microscopic concentrations in air or water samples, which creates anxiety in the public, especially because it vastly exceeds the regulatory capacity to establish safe levels of these contaminants. Public perceptions of risk are thus skewed, leading to a situation whereby politically unacceptable risks get funding priority over real ones. Finally, simply as a matter of human psychology, the public is predisposed to overrate risks that are imposed on them by external forces and underrate those deriving from activities in which they voluntarily engage. In such cases, scientific information is preempted.

Further complicating the public's response to environmental science is their view of scientists themselves, a view colored by the almost invariable portrayal of scientists in movies and on television as strange, antisocial, "nerdy," if not outright sinister types. This characterization of scientists may well derive originally from Frankenstein, the immortal embodiment of science gone mad. The image of this monstrous creation still lives in the subliminal mind of the public and thereby serves as a constant reminder of the dangers of unfettered science and the need to enlist its services only with the utmost caution.

The Precautionary Principle

Finally, it is only appropriate to conclude on a cautiously optimistic note. There is developing a protocol for addressing the most thorny issue in environmental policymaking—whether and how to address problems that have not yet manifested themselves or whose causes have not yet been clearly identified. As discussed earlier, those whose products and activities pose potential harm to public health and the environment have had some success in forestalling government action to address those threats by exploiting the scientific uncertainty we have been talking about. Specifically, they have argued that, in the absence of a definitive determination that those products or activities do in fact cause the alleged harm, public action to meet them would be, in many if not most cases, a ghost-chasing endeavor, squandering public and private funds and diverting attention from known and serious risks. To the environmental activists' warning that it is "better to be safe than sorry," the regulated community has retorted, "If it ain't broke, don't fix it."

The problem is a serious one, and there is solid ground for both arguments. Surely any number of alleged risks over the past several years have failed to materialize—those from electromagnetic fields come immediately to mind—and public policy ought not to be held hostage to the claims of any scientist, or any study, that purports to identify an environmental situation that merits a public response. That would indeed carry caution to an unreasonable extreme and have the effect of misallocating our finite fiscal and personnel resources. On the other hand, environmental interests have moved regulators to deal with some problems where they could credibly argue that waiting for scientific certainty to be established would preempt our ability to solve them at all or where the potential harm was irremediable. As the scope and consequences of environmental problems grow, and as our detection systems become ever more sophisticated, this problem has become all the more intractable.

The protocol that offers promise to help resolve this dilemma is the "precautionary principle." The precautionary principle, in concept, was born in Germany in the 1970s, was introduced in 1984 at The First International Conference on Protection of the North Sea, and was given formal expression at the 1992 United Nations Conference on Environment and Development, to which the United States was a signatory. It has progressively gained currency as the means to arbitrate the debate over when or whether to act in these kinds of situations. Although it has had only limited application in the United States, most notably in the 1990 Massachusetts Toxic Use Reduction Act, it constitutes an important component of over a dozen international treaties and laws, including the Montreal Protocol on Substances That Deplete the Ozone Layer (1987), the United Nations Framework Convention on Climate Change (1992), the Maastricht Treaty of the European Union (1994), and the Cartegena Protocol on Biosafety to the Convention on Biological Diversity (2000), which governs international trade in genetically modified foods.

The prevailing definition of the precautionary principle is as follows: "When an activity raises threats to the environment or human health, precautionary measures should be taken, even if some cause-and-effect relationships are not fully established scientifically. In this context, the proponent of an activity, rather than the public, should bear the burden of proof [of the safety of the activity]." This definition was agreed on by a group of activists, scientists, lawyers, legislators, scholars, and treaty negotiators convened at the Wingspread Conference Center in January 1998 by the Science and Environmental Health Network, a consortium of more than four dozen environmental organizations dedicated to promoting the use of science to protect the environment and public health. The conference was organized in response to the several different practices and principles floating around under the "precautionary" umbrella.

A few points about the principle ought to be noted at the outset. Its most salient feature is that it shifts the burden of proof from a prospective regulator to the party conducting the questionable activity or creating the potentially harmful product, a burden currently borne in the United States only by the pharmaceutical industry. But the "precautionary measures" it condones are not simple bans, or restrictions, or moratoria. Rather, they are more anticipatory than remedial in nature. The kinds of questions that the precautionary principle dictates are: How much contamination can be avoided? What are the alternatives to this product or activity, and are they safer? And is this activity even necessary? Ideally, questions regarding safety are raised and answered at the earliest possible stages. In essence, the precautionary principle is almost indistinguishable from "pollution prevention," another protocol that has achieved wide acceptance following the enactment of The Pollution Prevention Act of 1990, whose guiding rationale is that products and processes be designed so as to minimize the pollution they produce in the first place, rather than ensure that the risk they might pose is managed successfully at the disposal end. The precautionary principle is, thus, a more nuanced strategy than those that have traditionally been adopted in response to alleged potential threats, where the more conventional regulatory tools—particularly risk assessment—have been employed. Here, the purpose is not to decide how much risk is acceptable, given the benefits, but whether the risks themselves can be averted.

As noted earlier, the precautionary principle has been embraced more warmly in Europe than in the United States, notwithstanding the imprimatur of the President's Council on Sustainable Development and the American Public Health Association. But signs are that it still faces political opposition from the same forces that insist on scientific certainty as a precondition to action. The principal argument made by opponents is that it will be used as a pretext for blocking new technologies and restricting trade rather than as protection from potentially dangerous substances, or, to put it another way, that economic, not health, risks are its true targets. They allege that countries and states will exploit it to close markets to imports that compete with local industries and that it will be yet another, albeit superficially laudable, form of protectionism.

Still, the principle offers much promise. The political pressure from the international community, which is substantially more sensitive to and fearful of environmental threats than are Americans, is almost sure to mount. How exactly it will play out in this country, however, will be the result of yet other battles between competing interest groups and not the consensus of a disinterested body of scientific experts.

CENTRAL IDEAS

Although everyone believes that environmental policies ought to be guided by sound science, the scientific process and public policymaking are fundamentally incompatible. That does not stop the combatants in the policy wars from exploiting science for their own ends. The principal obstacle to a decisive role for science in policymaking is the tentativeness and uncertainty of its conclusions. But recently, a concept called the precautionary principle addresses precisely this problem.

Chapter 6

State and Local Governments: The "Other" Interest Groups

> This agreement [wherein Virginia Electric agrees to cut emissions to avoid suits by NY] is a critical affirmation of the role that states can play in not only environmental enforcement, but also the broader range of enforcement issues where states have compelling interests that are not always pursued by other levels of government.
>
> Eliot L. Spitzer

> The mobilization of local communities in defense of their space, against intrusion of undesirable uses, constitutes the fastest-growing form of environmental action, and the one that most directly links people's most immediate concerns to broader issues of environmental deterioration.
>
> Manuel Castells

Earlier, we looked at the ways in which Congress, the federal government's lawmaking body, undertakes the business of making environmental policy. We noted that Congress, by reason of its diverse representation, does not—indeed cannot—set down policies in autocratic fashion, but rather must acknowledge and reconcile a wide variety of interests to achieve the consensus necessary to make law. We also noted that this process is played out on a very public stage. So comprehensively do national affairs dominate media coverage that in the minds of most people, Congress and "government" are synonymous, and all other agencies and instrumentalities are its all but invisible handmaidens. The situation is not likely to change soon. A 1998 report underwritten by the Project on the State of the American Newspaper documents diminishing coverage of state government affairs and a decline in staffing at the statehouse level in more than half the states.

Ironically, this trend runs counter to an evolving and increasingly significant role for state and local governments in national environmental policymaking. It is, of course, well known that they implement and enforce national policies in a number of areas, a function that has been delegated to them over the past several decades, as we shall see. Aside from serving these administrative functions, or more accurately, largely by virtue of how and to what extent they are expected to carry out these functions, however, smaller units of government have compelling concerns with the nature of those federal policies themselves and thus feel compelled to participate in their formulation. Though they are generally represented by their congressional delegations, the particular problems of individual states and municipalities often are ignored or sacrificed in the welter of issues and forces that characterize national political discourse. For these reasons, it has been important for state and local governments to play an increasingly aggressive advocacy role in the formulation of environmental policy. Those who would understand the politics of environmental policymaking need to appreciate the nature and various forms of that participation.

State Governments and Federal Policy

For most of the twentieth century, responsibility for protecting the environment and for addressing ecological insults to it was borne almost exclusively by state and local governments and state courts. Environmental issues were regarded generally as local issues, and the federal presence in such matters was meager. Individuals who claimed to be harmed by irresponsible environmental behavior had to seek redress in court and overcome difficult burdens of proof. But in the 1970s, a flood of national legislation established the federal government as the dominant player in environmental protection. Laws regulating water quality, pesticides, noise, coastal management, endangered species protection, toxic substances, waste management and air quality were passed, and the EPA was charged with establishing minimum national standards in most of these areas. But Congress delegated to the states and municipalities responsibility for implementing and enforcing most of them and provided much, though not all, of the funding necessary to do so. Such an arrangement made a certain amount of practical sense, of course. States and local governments, being closer to the problems, have a better grasp of their sources and of the available options for addressing them. They can thus tackle them in the most efficient and effective way and in a manner that minimizes adverse economic and social consequences, which is especially important because most of these laws have quite different impacts on the various jurisdictions. It is, to that extent, a practical allocation of authority and action. Such power sharing, however, is not without its challenges—and its politics.

An unmistakable sign that more than just cooperative governance is involved was the concerted effort on the part of the Reagan administration to return power to the states when that administration assumed office in 1981. It was called the New Federalism, and it sought to restore what it regarded as the proper balance by shifting programs previously dominated by the federal government to state and local governments. The issues are different with respect to conservation and land use, so we will look at them separately later. In the area of pollution control, though, the so-called devolution of authority envisioned by President Reagan became very complicated, because the line between regulatory implementation and policy-making is complex and political.

Devolution of authority is a complex and political issue for a variety of reasons. First, politically conservative interests that would relax, if not eliminate, much environmental regulation have always preferred that authority be exercised at the most local level possible. State and local governments have traditionally lagged behind the federal government in financial resources and in expertise and are weaker "opponents." Most states have constitutional mandates to balance their budgets, so that they are often forced to prioritize their efforts rather than to fully satisfy federal requirements. In recent years, one of the rallying cries of states has been "federal mandate, federal pay," a call that has had significant resonance. In fact, a measure to require the federal government to fund new initiatives that would impose significant new costs on states was the only one of the environmental bills to be enacted into law in the Contract With America package on which the 104th Congress ran.

With respect to expertise, much the same is true. With a few notable exceptions, state governments generally do not have the levels of technical and scientific expertise in their administrative agencies that the federal government has. Additionally, the growing trend to impose term limits on state legislators has served to further dilute subject matter knowledge and experience. The situation is even more serious at the municipal level, where a great percentage of officials are laypeople acting in temporary, even voluntary, capacities. All these factors provide a sufficient incentive for anti-environmental interests to move environmental regulation down the government ladder.

Beyond budgetary and technical concerns, the reality is that state and local governments have particular and influential constituencies. Major industries, e.g., the pharmaceutical industry in New Jersey, the auto industry in Michigan, the oil industry in Texas, the tourism industry in Florida, and the agriculture industry in Iowa, exercise enormous political influence on their governments by virtue of the vast numbers of people they employ and the substantial contribution they make to their states' tax bases. However indifferent federal policymakers may be to these industrial interests, the states must not only be sensitive to the disproportionate impact a national policy may have on them, but also literally do battle on the federal legislative and regulatory fronts to minimize their adverse impacts. Good environmental policy may well be an important value for state governments and their residents, but economic viability is equally, if not more, important. That is why one so often finds business and state government as allies in a lobbying effort to shape or defeat, a federal policy.

If particular commercial and industrial interests can exert pressures on the states, it requires even less effort for them to influence municipalities, whose tax bases are even more fragile. Certain public functions, such as solid waste collection, wastewater treatment, drinking water quality, and growth management, are generally carried out by the local jurisdictions— counties and municipalities. These functions are strongly affected by federal legislation, but municipalities have little opportunity to affect policy in these areas. Only rarely do they try to work out their problems caused by federal mandates with Congress, but they do lobby their state governments, not only for help with Washington but also for direct assistance. Their situations are analogous to the state-federal relationship. Adjacent municipalities can adversely affect each other by their commercial or industrial activities or land use plans, the impetus for which are almost invariably economic benefits.

Quite aside from specific constituent interests, states and municipalities have more general interests to reconcile, because states are not even internally homogeneous. Pennsylvania, although a largely rural state, has two major urban pockets—Philadelphia and Pittsburgh. Colorado has vast skiing and park areas that attract tourists, but they are often at odds with Denver's urban areas, where air quality is a major problem. Much the same can be said with respect to the relationship between Illinois and Chicago. A similar, though more expansive, split is that between the densely populated inner cities of north New Jersey and the rural and agricultural areas of south New Jersey. Urban areas, as one might expect, have very different needs and are subject to different pressures than rural areas. The population shift toward urban areas documented by the 2000 Census will result in a corresponding shift in interests.

To deal with these kinds of problems, municipalities increasingly are lining up with sister municipalities in the same state to promote or protect their "territory" from towns with different industrial and population profiles. Issues such as the siting of potentially hazardous facilities, traditionally decided locally, are becoming increasingly rancorous, especially as federal efforts to promote economic opportunity in urban areas clashes with local concerns about healthy neighborhoods. Perhaps foreshadowing a new trend, more than a dozen of North Carolina's largest cities and towns have established the North Carolina Metropolitan Coalition, which hired a lobbyist to represent their concerns before the state legislature. The coalition will certainly be involved in a broad range of issues, but surely environmental matters will be among the more prominent.

State and local governments, then, must participate directly or indirectly in the federal policymaking process to protect the interest groups on which their own economic viability depends, as well as the more general interests they represent by virtue of their populations and geography. They often do this collectively, through national associations and organizations with similar problems. The National Council of State Legislatures, the Council of State Governments, the National Governors' Association, the United States Conference of Mayors, the National League of Cities, and the National Association of County Officials are among the more prominent of the alliances that they have entered into to help them meet their special challenges. Most of these were formed relatively early in the twentieth century, essentially to facilitate information sharing and communication. Although not lobbies in the usual sense, these organizations now increasingly play advocacy roles as well and afford smaller units of government more clout in Washington. More clearly political are the permanent offices in Washington that many states have established. It is the mission of these offices to stay abreast of legislation, regulations, and other policy initiatives that may affect the state and maintain a vigilant presence so that prompt action to defend that state's interests can be undertaken before more politically complicated responses are required.

The increasing role of states in environmental policymaking, for all these reasons, has been dramatic. The conservatives' attack on environmental activism in general, and the EPA in particular, since the Reagan ascendancy has had its effect. Environmental policymaking has devolved to where the states now issue most of the permits, initiate most of the enforcement actions, and fund most of the programs. The EPA seems more willing, if not eager, to turn over decision making to the states, as the discussion of regulation in Chapter 3 suggests. They have effected a broad shift in policy from prescriptive standards, whereby the federal government dictates not only what standards are to be met but specifically how they will be met, to performance standards, whereby states have flexibility to satisfy the standards in the manner they see fit. There is increasing cooperation between the EPA and the states. Almost three dozen states have entered into National Environmental Performance Partnerships (NEPPs) with the EPA, wherein states are afforded wide latitude on how to meet federal goals. While the name for these partnerships may suggest that the state contribution is limited to the realm of how to implement federal mandates, in actuality states are making policy along the way. Cooperation is also evident in the brainstorming that is taking place between the EPA and an association that was formed in 1993 by state commissioners of environmental protection called the Environmental Council of the States (ECOS), negotiations that attempt to reconcile the parties' respective demands. Such new initiatives enable state and local governments to better accommodate their residents' interests without violating federal strictures, or at least modify them with federal blessing.

As states gain parity with the EPA, major regulated industries seem more willing to negotiate with them, sometimes acceding to higher permit fees for a streamlined permit process and more reasonable regulatory decisions. Over the past 10 to 15 years, then, states have become major players in the game, not just the implementers of policy imposed from above. Neither the EPA nor other federal agencies, however, are prepared to abdicate responsibility for environmental threats and conditions that cross state boundaries. That is, indeed, the salient issue of federalism. The cliché is that "pollution knows no political boundaries," so that while states and municipalities may attempt to meet the standards in ways most favorable to them at the expense of other jurisdictions downwind or downstream, the federal government must remain a vigilant protector of regional and national populations and ensure that no jurisdiction—state, county, or municipality—can improperly burden other jurisdictions in the course of cultivating its own interests. This situation has prompted any number of battles both on the floors of Congress and in the courts.

Other Roles

The balance between state and national interests and responsibilities in many areas is delicate and constantly shifting, but environmental issues pose especially thorny problems because so many of them involve phenomena that embrace or cross state or regional boundaries. After all, the Mississippi River flows from the Canadian border to the Gulf of Mexico, inevitably carrying pollutants with it, and Ohio River Valley air pollution migrates to the Northeast. Regional associations among states, such as the Coalition of Northeast Governors (CONEG), have been formed to anticipate such problems and address them collectively. Energy issues, as well as pollution, cross state boundaries. The various energy grids shared by regional states are differentially affected by national energy policy, as the California power crisis dramatically demonstrates, and federal energy policy on efficiency standards impacts states differently, the source of an effort by New York, California, and Connecticut to oppose newly proposed regulations on air conditioners.

Typical of these battles between state and federal jurisdictional authority was a debate that took place in 1998, when the National Governors' Association, at their annual meeting, passed a resolution demanding control over how and when their states would meet federal Clean Water Act standards. Their proposal provided that each state would develop its own program to meet the standards. State and local governments would then work out a watershed plan for each river and lake in their jurisdiction. The state programs would be subject to EPA approval, but the local plans would not, and the criteria for rejection of a state plan would be hard to meet. Both the EPA and environmental activist groups objected to the proposal, and the withdrawal of federal funding was, characteristically, wielded as a weapon to ensure that national goals were met. However, federal sanctions, though often brandished, are seldom if ever used because the political fallout from such actions would be counterproductive to all parties. Their threat nevertheless often enables contending parties within a state to reach an agreement before any federal penalty would be imposed.

But state interests are not always impeded by federal strictures. Sometimes states in the same region prod the federal government to adopt stronger regulations. States have acted in concert to coerce industry to help them meet—or exceed—federal standards. Perhaps the clearest example of this was the action by New York State to adopt strict state vehicle emissions standards that are comparable to those in California. Adoption of the standards, especially when Massachusetts agreed to do likewise, would, in effect, force the auto industry to do what federal legislators could not do politically—make stringent emissions standards national—since it would be impractical and uneconomical for manufacturers to make different cars for different regions. New York, California, and Massachusetts together make up more than 25% of the national automobile market.

More often, however, state interests are at odds with each other. New York State, for example, has had perennial difficulty meeting federal Clear Air Act standards due in substantial measure to the air pollution carried by prevailing winds toward the Northeast from other parts of the country. Emissions from coal-fired power plants in the Midwest and South undermine New York's efforts to comply with federal requirements and to minimize acid rain by imposing stringent regulations on their own industries and citizens. Because the commerce clause provisions limit federal action, states simply go head-to-head in court to protect their interests and meet their needs.

Conservation Issues

It is readily apparent how the ambient nature of pollution creates special problems for states and municipalities and the industries for which they serve as hosts, requiring them to become aggressive advocates for these interests in policymaking forums. It is less obvious, though just as true, that land use issues impose corresponding obligations on them. Though lands and the resources they contain remain in place, of course, the resources themselves are not, and cannot be regarded as, the private property of the jurisdiction in which they are located. They are, rather, national resources owned by the federal government for the use and enjoyment of present and future generations of Americans.

At the heart of this clash between state and federal interests is the fact that much, and in some western states most, of the land is owned by the federal government. In the 1800s, in an effort to settle the West, the federal government offered economic incentives to states and private interests to settle them. Some of these incentives took the form of extremely attractive leases to graze lands, extract minerals and natural gas, and log woodlands, but the federal government, perhaps overgenerous in these initiatives viewed in hindsight, has maintained some degree of control over the commercialization of these resources. This control has persistently rankled both the western states and their extractive industries, which have regarded the federal government as a cruel, if not illegitimate, landlord. More than a hundred years later, in the late 1970s and 1980s, an organized effort was made by western senators to wrest control of these lands from the federal government in what was to become known as the Sagebrush Rebellion. When their efforts fell short, they were pledged support from the Reagan administration in general and from Secretary of Interior James Watt in particular.

An interesting twist of irony was instrumental in the failure of this effort. While western senators and their constituents were trying to effect the transfer of federal lands from the control and regulation of agencies such as the Bureau of Land Management, the Fish and Wildlife Service, and the U.S. Forest Service to the states, the conservative Washington establishment was promoting the "privatization" of these lands, under fear that state governments were perfectly capable of exercising almost as much unreasonable authority as the federal government. The interests that stood to gain commercially from relaxation of federal control feared that privatization would mean that they would have to buy these lands in the open market and pay appropriate taxes on them. Since they were presently enjoying "sweetheart" deals under which they paid virtually nothing for the rights to graze, drill, mine and log, privatizing these lands might well be worse for them. Thus there developed a split in goals between the private-sector industries and the federal government that caused the whole effort to lose focus and founder. Still, perennial legislative and judicial initiatives are undertaken by the states and their industrial constituents to secure more freedom to com-modify what they regard as their rightful possession. Although conservation issues in other parts of the U.S. have shorter roots and are generally less con-tentious, activities as diverse as logging in Maine's forests, to sugar produc-tion in Florida's Everglades, to ethanol production in Iowa's cornfields find the states in active advocacy roles to promote, protect, and defend the inter-ests important to them and their constituents that may not otherwise be acknowledged or negotiated by the political system.

It is relevant to note that, at the close of the Clinton administration and the onset of the George W. Bush administration, state and local interests seem to be gaining political strength. Three events during this brief interface support this conclusion. First, the Clinton administration's FY 2000 budget includes an increase of 5% over the amount Congress approved for environ-mental aid to state and local governments. The monies are to be used for expanded grants to states for purchasing ecologically sensitive lands; to assist cities in reducing air pollutants associated with global warming; for tax cred-its to subsidize local bonds for environmental purposes; and for contribu-tions to state revolving funds for municipal wastewater treatment facilities. While these funds are, in and of themselves, insufficient to fundamentally change the political calculus, they do represent an effort to minimize the adverse impact on states of federal environmental mandates, thus making state and local governments less economically dependent on their private-sector constituents.

Second, President Bush appointed Gale Norton as his secretary of the interior. Norton is a protégé of James Watt, the Reagan administration's inte-rior secretary who, as noted earlier, was at the center of the Sagebrush Rebel-lion of two decades ago. Like Watt, she was an attorney for the Mountain States Legal Foundation, a legal advocacy group that fought against the land management policies of the federal government. Most significant, however, she has been an outspoken proponent of more local, state, and private involvement in crafting the nation's land use programs. Just how strong her views are can be seen from a 1996 speech in which she analogizes the strug-gles between states and Washington to the Cold War: "Just as free markets tri-umphed over Communism . . . we are in a time when we can be part of the intellectual battle that shifts power from Washington back to states and local communities." Clearly, states and local communities will have an uncompro-mising supporter in Washington as future environmental policies are debated.

Finally, in January 2001, the United States Supreme Court, in a landmark decision, limited the scope of the federal Clean Water Act. Specifically, it pre-cludes the federal government from regulating small, local bodies of water (even though, collectively, they represent 20% of the area that could fall under the jurisdiction of the act), arguing that such regulation would repre-sent a significant infringement on the states' traditional primary power over land and water use, and thus curtails to some extent Congress's right to regulate commerce between the states. Although narrowly tailored to the specific situation before them—a number of localities sought to build a landfill upon seasonal ponds used by migrating birds—it does represent the disposition of the present Court's majority to favor state and local regulation over federal regulation whenever possible.

Thus, state and local governments are now in a stronger advocacy position than perhaps they've ever been. With the courts, the executive branch, and a substantial percentage of the general population leaning in their direction, we can expect them to be increasingly vocal in pursuing the special interests they represent. Government in this respect is clearly not a monolith, but, like the private sector, a many-layered complex of interest groups competing for favorable policy treatment. And some of those interest groups are governments themselves.

CENTRAL IDEAS

State and local governments, generally thought to be merely extensions of the federal government, are significant interest groups themselves. Their economic viability as well as their environmental conditions depend on constituencies not always served by federal policies. Accordingly, they have lobbied Congress much as other interest groups do and have formed alliances and associations to protect themselves.

Chapter 7

The Shifting Tides of Environmental Advocacy

> Saving the earth has never seemed so important, or so confusing. Not only do we have to deal with holes in our ozone layer, toxic seepage in our homes, vanishing lemurs in our burning rain forests, and too damn many copies of "999 Simple Things You Can Do to Embrace the Planet," we have also got to hack our way through a jungle of environmental organizations that's thick and getting thicker.
>
> Bill Gifford

> Our aim is to change the discussion within the boardrooms of major corporations. . . . That's where we will win ultimately, not in government agencies or Congress. Our strategy is basically like plugging up the toilet—by stopping them from opening up new landfills, incinerators, deep-well injections systems, and hazardous waste sites.
>
> Lois Gibbs

The major organizations that constituted the backbone of the environmental movement for almost the whole of the twentieth century are names familiar to almost all Americans. With the founding of the Sierra Club in 1892, the National Audubon Society in 1905, the National Parks and Conservation Association in 1919, the Izaak Walton League of America in 1922, the Wilderness Society in 1935, the National Wildlife Federation in 1936, the Defenders of Wildlife in 1937, and the Conservation Foundation in 1938, the environmental establishment incrementally became a major political force in American society. Collectively, these groups have become virtually synonymous with what are generally referred to in the media as "the environmentalists." Individually or jointly, they have participated in virtually all the major battles that have been fought over environmental issues at least since the struggle over Hetch Hetchy in the wake of the San Francisco earthquake of 1906. In so doing, they have effectively defined environmentalism, and no discussion of environmental politics can even begin without acknowledging the value, nature, and extent of their involvement.

But *how* they have defined environmentalism and the constituencies that they have represented in their advocacy over that time provide important lessons in the history of environmental politics and a window on where environmental politics is going in the future. Moreover, the new groups that have grown up alongside the major organizations since approximately 1980—groups that have challenged the establishment for public attention and commitment—are causing the establishment to rethink its mission, carving out radically new directions for environmental policy, and forcing governments at all levels to review their policies and procedures.

Over roughly the first half of the twentieth century, the conservationist agenda of the major environmental organizations was clear and focused: to protect and preserve the natural resources of this country. They fostered the association of environmentalism with nature—with wilderness, with coastlines and lakes and streams, with mountains and canyons, with birds and wildlife, and with exotic flora and fauna. They did so in the service of a constituency of outdoor enthusiasts and travelers who were wealthy and politically savvy and whose overriding interest was in maintaining a clear line between the pristine and the polluted and fearful of losing something precious and irreplaceable to "civilization."

But as the environmental agenda slowly began to change, to focus increasingly on air and water pollution, on the generation of wastes and their irresponsible disposal, on pesticide applications and their effect on human health, and on the impacts of energy production and development, the major environmental organizations began to experience internal conflicts. They felt constrained to attract a wider membership base, adopt new political strategies, and expand their missions beyond their core conservationism. These changes were necessary to reflect the expanded environmental agenda that developed in the 1960s and 1970s. So, too, did the creation of sister organizations—the Environmental Defense Fund in 1967, the Natural Resources Defense Council in 1970, and the Earth Justice Legal Defense Fund (originally the Sierra Legal Defense Fund) in 1971 among others—to provide the litigation and the scientific and technical expertise required to enforce the scores of new laws and regulations passed by Congress and the administrative agencies. By the end of the century, literally thousands of new environmental groups of a whole new order sprang up all over the country. These organizational developments mirror the significantly altered policy posture that environmentalism has adopted over the past several decades.

The evolution began to occur at about mid-century, when the country "discovered" the environmental degradations associated with the burgeoning industrial activity that followed World War II. Though their memberships remained largely white, male, and upper class, and their goals accordingly preservationist, the national organizations began to take a more activist political stance. They lent their influence to the more general push for a quality environment, mainly through increased lobbying of Congress. Many established offices in Washington and testified on the full range of issues, some unrelated, or only peripherally related, to conservation. They also took on a public education role and, increasingly, filed court suits against both private-sector violators of environmental laws or regulations and governments who failed to fully enforce them.

David Brower and the Sierra Club

We can see this evolution in microcosm—its accomplishments, its internal strains, and a foreshadowing of its new directions—in the work of the seminal environmentalist of the century, David Brower, who served as executive director of the Sierra Club from 1952 to 1969. A successor and ideological twin of John Muir, who founded the Sierra Club, Brower dedicated his early career to making it a force in forging environmental policy. He maintained pursuit of its historical mission in a series of successful battles to preserve Dinosaur Monument and Point Reyes, establish North Cascades and Redwoods National Parks, block the construction of two hydroelectric dams in the Grand Canyon, and, most notably, promote the enactment of the Wilderness Act in 1964. The tactics he employed in these efforts, however, were distinctly more political than the Sierra Club's old guard, and the IRS, were comfortable with. His unsuccessful attempt to halt the construction of the Diablo Canyon nuclear reactor in California was only the last in a series of aggressive political activities that caused the club to lose its tax-exempt status and to accept his resignation.

During Brower's tenure, the Sierra Club increased its membership more than ten-fold, making its constituency obviously broader and more diverse, as well as more politically active, and taking the first steps, however tentative, toward an expanded environmental presence. Brower also refined two other political tactics that would become common in succeeding years. He made direct personal appeals to influential government officials with jurisdiction over the matters he was concerned about, and he raised to a fine art the persuasive power of the coffee-table-size photographic depictions of the natural beauty and wildlife that would be violated, if not threatened with extinction, if proposed developments went forward. Conceived and utilized as inanimate lobbyists, such books became popular among the general public as well and reinforced the conservationist message. Bower also used another technique that now has become a staple of environmental politics—full-page letters to the public in the *New York Times*, making the case and encouraging readers to petition their representatives in Congress for help.

After leaving the Sierra Club, Brower continued to anticipate where the viability of the environmental movement lay. He became more conscious of the relationship between conservation and urban issues such as traffic, polluted water, and consumerism. He also involved himself in the broader ecological crisis. For example, he fought (albeit unsuccessfully) federal approval to lay the Alaska pipeline at the risk of damage to the surrounding landscape and wildlife habitat and the development of the supersonic transport aircraft battles, which would cause, among other consequences, a significant noise impact. Friends of the Earth and the Earth Island Institute, both of which he founded, took on problems that had global dimensions, issues such as atomic arsenals and nuclear proliferation. His instinct that environmentalism would have to embrace concerns related to the arms race and social justice was uncannily prophetic, for that was precisely where environmentalism was going, as we shall see. In the last years of his life, Brower helped organize the Alliance for Sustainable Jobs and the Environment, bringing together environmentalists and steelworkers in what has become one of a new breed of environmental organization.

Growing Politicization of Environmental Groups

Brower's career serves as a template for environmental organizational development after mid-century. While the major groups continued to pursue their own agendas, consistent with their respective missions and memberships, they also became increasingly involved in the growing political battles over environmental and conservationist issues, or, to put it more precisely, to draw the connection between them. That involvement, however, was largely through the traditional channels—active participation in the legislative and regulatory processes—in a manner that was polite and civil. They sought to influence policy by influencing the policymakers directly, as befit the disposition of their traditional core constituencies. But their newly emergent political activism spawned new allies—the Environmental Defense Fund in 1967 and the Natural Resources Defense Council in 1970.

These new offshoots differed from the mainstream groups in several ways. First, though they eventually sought out and secured members, their funding initially came principally from foundations and a handful of wealthy contributors. Second, they focused on litigation and, later, on negotiated agreements with business and industry. Last, they apportioned the agenda, each taking principal responsibility for certain issues. Their freedom to pick issues and tackle them as they saw fit was made possible in large measure by their relatively less restrictive financial support.

During the 1960s and 1970s, the major organizations and their more litigious brethren, though they took different tacks, assumed the responsibility to make the environmental gains secured in the governmental arena stick. During this period they represented the broad public interest against the encroachments of the corporate polluters. These were the decades during which their reputation as "the environmentalists" was solidified and the time when environmentalism became an accepted national value. The environmental community and the broad general public were essentially in sync.

But things were to change significantly in the 1980s. The principal event was the election of Ronald Reagan, whose agenda targeted environmental regulation directly and promoted the mining, logging, drilling, and grazing of land, as well as increased state ownership of federal lands. So threatening was the prospect of his successful implementation of this agenda that the public responded enthusiastically to the environmental organizations' calls for help, swelling their membership rolls dramatically. During the 1970s, their growth had been modest, and in some cases flat; the exception was the National Audubon Society, which more than tripled its membership, from a little over 100,000 to almost 400,000. During the 1980s, however, memberships made quantum leaps. The Sierra Club, the National Parks and Conservation Association, the Environmental Defense Fund, and the Natural Resources Defense Council each more than tripled in size, and the Wilderness Society increased more than tenfold.

Such numbers reflected the growing public interest in environmental matters to be sure, but they also reflected both a new political acumen regarding these organizations and a sincere public concern that the Reagan administration would roll back environmental gains of the past decade. In particular, Interior Secretary James Watt's pursuit of land use policies that encouraged the commercial exploitation of many of the natural areas that these groups regarded as veritably sacred was probably the best fund-raising tool, as Americans obviously looked to these groups for protection from an unfriendly executive branch.

Most of the environmental cutbacks contemplated by the Reagan administration were forestalled by a strengthened alliance between the major environmental organizations and the environmentally progressive members of Congress and the federal agencies. Nevertheless, two other factors began to have a countervailing effect. First, the level of political involvement engaged in by the mainstream organizations demanded new levels of funding well beyond what membership dues could provide. Increasingly, they had to seek new sources of financial support. One has been foundations, most of which are genuinely philanthropic and politically neutral, but some of which have their own political agendas. A second source of revenue is large corporate donors, which have a variety of motives to share their wealth with the "enemy," some simply to build environmentally friendly images for themselves in the minds of a public for whom environmentalism had become next to godliness, but many, no doubt, to round off the adversarial edge between them. Still another source of revenue is national mass mailings, in which requests for money are accompanied by vivid descriptions of the violence that would be done to one or another of our most precious natural resources or endangered species or to human health by a present condition or policy that needs a champion. More recently, as the influence of the public on policymakers has become unmistakable, the typical single-page format of such mass mailings has taken on a three-part character: a brief lobbying statement about some problem or threat that the organization will fight to alleviate; a prepared form letter to the recipient's congressional representatives or to the president, entreating the policymaker to address the issue legislatively; and finally the obligatory request for a donation.

The increasing need for funding and the growing reliance on business interests to supply that funding have, in the minds of some observers, compromised the integrity of the major organizations. Some have even gone so far as to suggest that the environmental lobby, except in its broad parameters, is now almost indistinguishable from its business counterparts. For confirmation, they look not only at their plush and commodious offices in Washington, but also to the business-friendly compromises that they have too often made to address problems. Specifically they point to accommodations environmental organizations have made with polluting industries. One example is the pollution tax credit, wherein businesses are provided a tax incentive to reduce pollution rather than face penalties for exceeding established standards. Another is the so-called bubble-policy, which treats an industrial facility as one large emission source that must satisfy a cumulative standard rather than as, under traditional regulation, a number of separate sources, each of which must meet standards. Another contentious accommodation is the incorporation of cost-benefit analysis, which introduces economic burdens as qualifying factors in setting emission limits. Less subtle are the "clubby" relationships between business and the organizations fostered by the fact that prominent corporate officials are not infrequently on the boards of these organizations, which surely facilitates the securing of grants and other financial help but just as surely calls into question the ideological purity of the organizations' commitments.

Another factor that has affected the mainstream's political posture is the concentration of environmental groups' offices and activities in Washington and New York and their focus on national issues, often at the expense of local ones. Local chapters of the national organizations have, by and large, decreased in number and significance as power moved to the

Northeast. This relocation trend, together with their effort to work pragmatically from within the system rather than from without, has had substantial consequences as well as some achievements. One that attracted much media attention was an agreement negotiated by the Environmental Defense Fund with McDonald's, which resulted in significant reduction of Styrofoam waste.

The Rise of Environmental Justice

Whether certain segments of the general public felt abandoned by the organizations or whether they felt comfortable that the overall environment was adequately protected by them is a matter of speculation. What is clear is that concurrent with the growth and corporatization of the national environmental organizations in the 1980s there appeared the budding of another, quite different, movement. The early signs took the form of citizen action groups that assembled to address a particular concern or problem or to accomplish a single purpose. For example, some volunteer groups "adopted" a beach, a highway, or some other public tract, taking upon themselves the responsibility to keep it clear of litter and debris. Other groups—"friends of open space" or "citizens united to prevent urban sprawl"—keep watch over local planning and zoning boards to ensure that development is consistent with land use plans that protect the environment. There emerged ad hoc coalitions of local citizens who organized themselves to protect specific wetlands, beach areas, or woodlands. These community activists focused less on policy and more on addressing specific problems or threats that affected their daily lives in their own neighborhoods. It should be noted, however, that these early groups were concerned principally with amenities—like litter—rather than significant environmental threats.

While a small segment of civilian activists were engaging in efforts to save their beaches, their lawns, and their property values, a far more significant, militant, far-reaching, and diverse movement was taking hold. It shared with these local groups an interest in the environment of their daily lives and a commitment to action rather than policymaking, but there the similarities end. This new movement concerned itself not with amenities but with health—with exposure to toxic pollution and radiation in their workplaces and communities.

The constituents of this movement, unlike those of the national groups and even the "friends" groups, occupy the middle or lower economic strata and comprise disproportionately people of color, aggrieved minorities, and women. They see themselves as victims of environmental degradation rather than as guardians of environmental integrity. To put it more bluntly, they feel "dumped on," with the "dumpers" as often governments as corporate institutions. They are angry, because, as they see it, their civil rights as well as their homes have been violated.

Although over time some organizational structures have developed to provide education, technical assistance, and networking capacity to the individual groups, the movement is fundamentally decentralized, disparate, and grassroots in character. The individual groups see it as their mission to take control of their own destinies rather than trust government, or even supposedly environmentally friendly organizations and institutions, to protect their homes, neighborhoods, and workplaces. They distrust, as well, the formal processes of lawmaking, and thus do not lobby or seek financial support from special interest groups.

It is always hard to determine just when and how grassroots movements begin, but it's safe to say that there was a latent, if amorphous, concern about the health dangers posed by toxic chemicals that threatened Americans in many quarters. Reports about the dangers of Agent Orange, PCBs, radioactive waste, and dioxin had been reported in the press time and again during the 1970s. Some highly publicized events—the discovery of dioxin at Times Beach, the children's cancer cluster at Woburn, and the malfunction of the nuclear power plant at Three Mile Island—only reinforced public unease and made these threats more than hypothetical matters. The discovery of hazardous chemicals in the Love Canal neighborhood of Niagara Falls, NY served as the critical spark that ignited a firestorm of protest.

Though by no means unique, the situation at Love Canal has come to be regarded as the pivotal event in the birth of this new citizens movement. This is attributable in large part to Lois Gibbs, one of the resident victims, who channeled her outrage first into a successful effort to require the government to relocate the hundreds of families potentially affected and then into the founding of the Citizens Clearinghouse for Hazardous Waste (CCHW). CCHW galvanized the existing unease in hundreds of communities nationwide into a genuine environmental protest movement. In the almost 20 years since its creation, the CCHW has grown into a substantially larger network of local groups, now numbering more than 8,000, with a new name—the Center for Health, Environment and Justice. The new name is not just window dressing; rather it is emblematic of the enlarged set of values underpinning the movement. Though initiated to help respond effectively to specific, local concerns, the myriad efforts are animated by a sense that not only the environment but also social justice is being threatened. Marrying good environmental policy and social justice, however, has proved to be a daunting task indeed.

Before getting to that stage, however, it is instructive to see how the federal government and the national environmental organizations are responding to this "interloper" and the problems posed in trying to integrate them into environmental policymaking. As noted, this new movement—the "environmental justice" movement—has gained momentum and influence over the past two decades. In many ways, it embodies and reflects much of what has happened to mainstream environmentalism in recent years, and most major environmental organizations have embraced it and welcomed its constituents into their tent. For a number of complex reasons, though, a substantial segment of the environmental justice community has not only insisted on its own identity, but has actively opposed some of the initiatives of the EPA and other traditional environmental forces. That tension grows out of the related, but by no means congruent interests of social justice and environmentalism, a conflict that serves as a veritable paradigm of environmental politics as it stands at the onset of the twenty-first century.

As a clear demonstration of the social and political success and influence that the environmental justice movement had attained, on February 11, 1994, President Clinton issued Executive Order no. 12898 directing each federal agency to "make achieving environmental justice part of its mission by identifying and addressing . . . disproportionately high and adverse human health or environmental effects of its programs, policies, and activities on minority populations and low-income populations." The order was extraordinarily broad and deep, extending specifically not only to the United States, but also to its territories and possessions, including Puerto Rico and the Commonwealth of the Mariana Islands. Furthermore, the Interagency Working Group on Environmental Justice created by the order comprised representatives from the EPA, from the Departments of Defense, Health and Human Services, Housing and Urban Development, Labor, Agriculture, Transportation, Justice, Interior, Commerce, and Energy, and from the Offices of Management and Budget, Science and Technology Policy, the Assistants to the President for Environmental Policy and for Domestic Policy, and from the National Economic Council and the Council of Economic Advisors. The group's delegated responsibilities were to assist these federal departments and agencies in developing criteria to identify the disadvantaged populations, coordinate research and data collection, and, in general, help them prepare, administer, and enforce the individual strategies that the order required each to develop to address the problem.

The problem, as the order suggests, is that people of color and those in the lower economic strata of our society are subject to a disproportionate health risk from environmental pollutants and conditions associated with the industrial plants and waste facilities that are too commonly sited in their neighborhoods, as well as from activities and circumstances to which they as a class are unreasonably subject. The evidence for this conclusion was collected at first anecdotally and later more systematically but culminated in a definitive national study, *Toxic Waste and Race in the United States*, conducted by Charles Lee and published in 1987. The study, in lending substance to the claims of the growing environmental justice movement, represented a watershed in its supporters' efforts to get public recognition of these inequities and government action to redress and compensate them.

The next couple of years saw renewed attention to the issue from Congress, government agencies, and scholars at conferences nationwide. In July 1990, EPA Administrator William Reilly established the Environmental Equity Workgroup to review the evidence of these alleged inequities and report to him. The group's confirmation of the charge led in turn to the formation of an Environmental Equity Office to provide technical assistance, outreach, and communication to the movement. By 1991, the First National People of Color Environmental Leadership Summit was held in Washington, D.C., at which its principles were first codified. In October 1992, the first significant efforts on the part of the EPA to actively address the problem occurred through a series of memos circulated to staff. Meanwhile, the reinvigorated grassroots activism of the supporters resulted in the formation of an Environmental Justice Transition Group, a diverse consortium of interests whose recommendations to President Clinton's transition team culminated in the executive order.

This order represents a major milestone in the environmental politics of this country—a pioneering initiative on the part of the federal government to enlist its combined resources in a systematic, proactive effort to protect and promote the health and environmental circumstances of a particular segment of the population, a group of interests related by their mutual victimization, if you will. Because that interest group is defined principally by its racial and ethnic backgrounds and income, the order effectively brings together the environmental movement and the civil rights movement that immediately preceded it and from which it drew many of its first recruits.

The movement's roots are to be found, unsurprisingly, in the Civil Rights Act of 1964 and particularly in Title VI of that act, which accords minorities equal protection under the law and specifically prohibits discrimination against people based on their race, color, or national origin or denies them benefits from, or participation in, any government activity that receives federal funding. That fundamental principle was extended to environmental programs in 1973 by EPA's issuance of a regulation to implement Title VI in their programs and again in 1984 by another regulation prohibiting any recipient of EPA financial assistance from undertaking any policy that has a discriminatory effect under the Civil Rights Act. These legal authorities were specifically invoked by President Clinton in his executive order.

There is, then, a clear statutory and regulatory foundation on which to base action to redress environmental "racism," as it has been called, and to empower low-income, minority, and ethnic peoples to participate in policy decisions that may, if made in the traditional institutional ways, deprive them of their basic human rights. In recent years, both the executive branch and

Principles of Environmental Justice
Adopted October 27, 1991
1st People of Color Environmental Leadership Summit
Washington, D.C.

Preamble

WE, THE PEOPLE OF COLOR, gathered together at this multinational People of Color Environmental Leadership Summit; to begin to build a national and international movement of all peoples of color to fight the destruction and taking of our lands and communities, do hereby re-establish our spiritual interdependence to the sacredness of our Mother Earth; to respect and celebrate each of our cultures, languages and beliefs about the natural world and our roles in healing ourselves; to insure environmental justice; to promote economic alternatives which would contribute to the development of environmentally safe livelihoods; and to secure our political, economic and cultural liberation that has been denied for over 500 years of colonization and oppression, resulting in the poisoning of our communities and land and the genocide of our peoples, do affirm and adopt these Principles of Environmental Justice:

1. Environmental justice affirms the sacredness of Mother Earth, ecological unity and the interdependence of all species, and the right to be free from ecological destruction.

2. Environmental justice demands that public policy be based on mutual respect and justice for all peoples, free from any form of discrimination or bias.

3. Environmental justice mandates the right to ethical, balanced and responsible uses of land and renewable resources in the interest of a sustainable planet for humans and other living things.

4. Environmental justice calls for universal protection from nuclear testing, extraction, production and disposal of toxic/hazardous wastes and poisons and nuclear testing that threaten the fundamental right to clean air, land, water, and food.

5. Environmental justice affirms the fundamental right to political, economic, cultural and environmental self-determination of all peoples.

6. Environmental justice demands the cessation of the production of all toxins, hazardous wastes, and radioactive materials, and that all past and current producers be held strictly accountable to the people for detoxification and the containment at the point of production.

7. Environmental justice demands the right to participate as equal partners at every level of decision-making, including needs assessment, planning, implementation, enforcement and evaluation.

8. Environmental justice affirms the right of all workers to a safe and healthy work environment without being forced to choose between an unsafe livelihood and unemployment. It also affirms the right of those who work at home to be free from environmental hazards.

9. Environmental justice protects the right of victims of environmental injustice to receive full compensation and reparations for damages as well as quality health care.

10. Environmental justice considers governmental acts of environmental injustice a violation of international law, the Universal Declaration On Human Rights, and the United Nations Convention of Genocide.

11. Environmental justice must recognize a special legal and natural relationship of Native Peoples to the U.S. government through treaties, agreements, compacts, and covenants affirming sovereignty and self-determination.

12. Environmental justice affirms the need for urban and rural ecological policies to clean up and rebuild our cities and rural areas in balance with nature, honoring the cultural integrity of all our communities, and providing fair access for all to the full range of resources.

13. Environmental justice call for the strict enforcement of principles of informed consent, and a halt to the testing of experimental reproductive and medical procedures and vaccinations on people of color.

14. Environmental justice opposes the destructive operations of multi-national corporations.

15. Environmental justice opposes military occupation, repression and exploitation of lands, peoples and cultures and other life forms.

16. Environmental justice calls for the education of present and future generations which emphasizes social and environmental issues, based on our experience and an appreciation of our diverse cultural perspectives.

17. Environmental justice requires that we, as individuals, make personal and consumer choices to consume as little of Mother Earth's resources and to produce as little waste as possible; and to make the conscious decision to challenge and reprioritize our lifestyles to insure the health of the natural world for present and future generations.

the EPA have embarked on such an effort. The environmental establishment has been generally supportive, at least recently.

Yet both the government and the environmental advocacy groups have found that initiatives to promote a safe and healthy environment for the targeted populations identified in the order are complicated by the mandates of other federal programs and policies and by a long and complex social and cultural history. They are also presenting new challenges not only for the EPA, but for environmental policymaking in general. Let's address these one at a time.

The effective implementation of the order in any number of cases has met with strong opposition, and from precisely those whose interests it was specifically designed to protect. There can be no justification for disproportionately subjecting any group of citizens to the ill effects of environmental degradation, much less those who are politically least able to defend themselves. That is the unimpeachable basis for the scores of protest marches, policy papers and books, and conferences condemning the situation. Yet officials in local and state governments, much of the business community, and a not insignificant percentage of the allegedly aggrieved populations themselves see those initiatives as impeding federal programs such as those establishing "empowerment" or "enterprise zones" that offer financial incentives to businesses to bring economic development into poverty areas and create jobs for minorities and the poor.

Environmental justice principles are also at odds with the so-called "brownfields" program, whereby the federal government is promoting the cleanup and reuse of abandoned industrial sites in urban areas rather than siting new plants in pristine areas, again to advance the redevelopment of blighted areas and create jobs for those most in need. Environmental justice advocates argue forcibly that clean jobs can be created in minority communities as well and that one should not have to make the choice between health and work. Harry Alford, president and CEO of the National Black Chamber of Commerce, however, characterized the effect of environmental justice policy as "preserving the economic blight of urban communities." Years earlier, urbanologist Anthony Downs put it more bluntly: "The elite's environmental deterioration is often the common man's standard of living."

The EPA is thus legally mandated to concurrently carry out two policies that have differing, and in some cases competing, effects on the same population. More particularly, the populations whose environmentally acceptable homes, neighborhoods, and workplaces are the goal of the environmental justice policy often need to pursue a more pressing and immediate goal—jobs and economic advancement—which environmental interests sometimes impede. This clash of interests has pitted government agency against government agency, social justice group against social justice group, and minority citizen against minority citizen. The mainstream environmental organizations have been sympathetic to the plight of the poor, to be sure, but they, too, are torn between their central mission and some of the unintended effects of its blind pursuit.

In a larger social and cultural context, the strife occasioned by the ascendant environmental justice movement is only a recent manifestation of a broader ideological clash between environmentalism and social justice that goes back more than a quarter century. For reasons suggested earlier in this chapter, environmentalists have historically been characterized by black activists, blue-collar workers, and political scientists and sociologists as a social elite—predominantly white, upper- or upper-middle-class professional suburbanites whose basic living needs have long since been satisfied and who can now indulge in a privileged lifestyle that values environmental quality without regard to the price it may exact on the economy, government, or other human welfare concerns. Its harshest critics have even charged environmentalism with bigotry and racism in its attempt to protect the lifestyle of the comfortable at the expense of the poor and underprivileged. These critics point to environmentalists' support of such policies as population control, open space preservation, large-lot zoning and other growth-control strategies as evidence of a thinly veiled effort to impede the upward mobility of the have-nots and thus protect their own homes, neighborhoods, and recreational areas from the encroachment of urban pressures. However environmentally desirable, such policies have the effect of segregating economic classes and minimizing, if not eliminating, industrial activity in the most affluent areas.

Indeed, it was in such a context that the label NIMBY (not in my back yard) was coined to refer to those suburbanites who oppose the siting of LULUs (locally undesirable land uses) in their neighborhoods. Their political and economic might often effectively stymie the location of industrial plants and waste facilities in their upscale surroundings. Although the motivation for their protests is often the protection of property values, too many of their efforts aim to improve quality of life by promoting clean air, clean water, and responsible toxic management for them to be written off as self-serving.

Given the broad political profile and voting history of mainstream environmentalists, it would be difficult to label them as deliberately antipoor and racist, but we have seen that environmentalism was, at least in its origins, the province of the more well-to-do and politically active and effective. Statistics on membership in environmental organizations; on visitations to parks, forests, and other such recreational areas; and on participation in activities such as hunting, fishing, backpacking, birdwatching, and hiking all attest to an interest group served by a conservation agenda and distinctly unrelated, if not indifferent, to urban needs and concerns. It is an identity that has dogged the movement.

These kinds of charges have persistently stung the environmental community, who rightly insist that all citizens, regardless of their station in life, are equal beneficiaries of clean water, clean air, open space, parks, forests, and wilderness areas and thus take comfort that they are serving a solemn mission. Even more galling to them is that the business and construction interests, their principal political adversaries, have exploited the "elitist" identity and have proudly proclaimed—with varying degrees of integrity and effectiveness—that their industrial and commercial activities in fact serve the genuine interests of "people," instead of animals, trees, and flowers, and that the so-called altruism of the environmental groups is narrow and self-interested. In the present dispute over the application of environmental justice policy, the not uncommon alliance of business and environmental justice protesters gives credence to the claim.

Against this historic backdrop, one can see that the clash over the implementation of the executive order reflects a longstanding and fundamental tension between the interests served by programs to promote environmental quality and those served by programs that advance social justice. It could, and hopefully will, eventually lead to better environmental policies for everyone. That is, after all, its intent. Much of what the environmental justice movement has brought to environmental advocacy is desirable. It has redefined citizen empowerment, demanded the wider public distribution of pollution data, strengthened the push for pollution prevention, influenced consumer and business behavior, and, perhaps most important of all, broadened and democratized the environmental constituency.

On the other hand, the environmental justice movement poses formidable problems for environmental policymaking and for the national organizations. First, it effectively asks for environmental policy to redress other social and public health inequities, a lot of baggage for essentially siting decisions to carry. It interposes a new element in a whole range of policy options that emerged from the EPA's reinvention and thus creates as much policy confusion as it may illuminate. Further, it complicates the relationship between the federal government, which is charged with carrying out this new mandate, and state and local governments, in whose authority reside many of the siting decisions. Perhaps most threatening of all, it challenges the ability of the governmental establishment itself to solve these problems, an establishment that mainstream organizations have worked with for years and in which they retain considerable stake. It also obscures the very real accomplishments of the mainstream organizations, accomplishments that are attributable to their spirit of practical compromise. Environmental justice groups have taken a role in the political drama, but they must share the stage with an establishment that continues to give environmentalism standing among the general public and in Congress, particularly among lawmakers who are uncomfortable with renegade troops. Further, the establishment continues to provide valuable funding for communication and research services that are indispensable to the success of everyone's efforts. Tensions remain between the camps, to be sure. How they are resolved will be determined by how the environmental establishment accommodates its new constituency and the extent to which the grassroots alliances acknowledge and embrace the critical role played by the mainstream groups to their own existence. Their reconciliation—or split—will foreshadow the direction, and perhaps the viability, of the environmental movement itself in the years ahead.

The Critique of Culture

If the 1980s saw the national environmental organizations and the nascent environmental justice movement each claiming a segment of the population and, in their respective ways, trying to shape policy to satisfy their respective constituents, the 1990s witnessed a tilt away from traditional environmental advocacy and toward a diversification of causes. The membership gains enjoyed by the mainstream organizations during the 1980s, almost surely in response to concerns about the policies of the Reagan administration, eroded in the 1990s. Most observers attribute this loss to a revived level of comfort about the environmental stewardship expected from an environmentally friendly administration, particularly Vice President Gore, long an outspoken environmental advocate and author of *Earth in the Balance*, a book setting forth an extraordinarily activist agenda.

But the inroads made by environmental justice reformers may also have started to take their toll. In addition to questioning the will, if not the capacity, of the mainstream organizations to attend to the specific problems of specific people in specific places in deference to national policy concerns, the environmental justice reformers, by introducing social and economic issues into the discussion, opened the national organizations to criticism from the other quarters and paved the way for another redefinition of environmentalism. In the 1990s, environmentalism became a cause in itself as well as an element of a broad panoply of concerns.

Once social justice became linked to environmentalism, labor, peace, arms control, population control, immigration policy, globalization, industrial agriculture, poverty, and a host of other similar issues were commingled with it, again as Brower anticipated. The critique of our deteriorating environmental condition became a critique of the culture at large, both nationally and globally. This gave rise to a number of alliances between previously separate groups, alliances that began to lobby not so much as interest groups but as global saviors.

Two specific cases in point are illustrative. Zero Population Growth (ZPG) is a 30-year-old organization dedicated to slowing national and global population growth on the grounds that overpopulation is a root cause of, among other things, hunger, water scarcity, global warming, urban sprawl, and poverty. In 2000, ZPG sent out a national mailer, which set forth ZPG's mission to seek "sustainable balance between the Earth's people and its resources" by pursuing population control through a broad agenda of public education, legislation, and financial support for family planning. But the mailer went beyond its essential character as a lobbying statement, with the obligatory contribution envelope enclosed. It included an explicit attack on the refusal of Wal-Mart, one of the nation's largest retailers, to market a newly approved emergency contraceptive and a form letter to Wal-Mart's president, to be signed by the recipient, expressing outrage at a policy it characterizes as "an expression of contempt for the health needs of women." In attributing several adverse environmental and social problems to their organization's more narrow focus, overpopulation, and in soliciting the help of citizens in targeting directly a business practice that they see as symptomatic of the larger problems facing society, these alliances of heterogeneous groups are emblematic of the new stance that environmental lobbying is taking.

Another new organizational form of environmental advocacy is that practiced by the Turning Point Project. The Turning Point Project exists only on paper, or, more accurately, on a website turned interest group. Specifically, it is a nonprofit coalition formed in 1999 to design and produce a series of educational advertisements, for publication in the *New York Times*, on major issues of the new millennium on the quality of life on Earth. The more than 50 organizations that have signed on to one or another of the published advertisements are bound together by their common advocacy of democratized, localized, and ecologically sustainable alternatives to current practices. Established environmental organizations and institutions such as the Earth Island Institute, the Silicon Valley Toxics Coalition, Food and Water, the Ecologist, the Nature Institute, and the Rainforest Action Network have been joined by dozens of other more broadly cultural, educational, governmental, and media watch organizations as various as the American Academy of Pediatrics, the Center for Media Education, the Foundation on Economic Trends, the National Governors Association, the National Parent Teacher Association, the Organic Consumers Association, the Humane Society, and the Center for Food and Safety. The ads in the series take on the allegedly adverse consequences of megatechnology, industrial agriculture, genetic engineering, economic globalization, and the loss of biodiversity, which they see as elements of a larger crisis driving the planet toward social, political, and environmental breakdown.

As different as ZPG and the Turning Point Project are in many respects, both betray some essential trends in environmental advocacy. Both see environmental issues as part of a larger matrix of social, political, and cultural concerns and thus approach their advocacy from a more holistic and global perspective. Both appeal directly to the general public by utilizing national mail solicitation and the pages of the *New York Times*, respectively, as their vehicles of communication, and both are activist in orientation. The evolution of the Citizens Clearinghouse for Hazardous Waste into the Center for Health, Environment and Justice spoken of earlier was a clear sign of a new direction, to be followed not only by those discussed here, but by hundreds, perhaps thousands, of hitherto unorthodox alliances. It is ironic that scores of rear guard actions by local protesters addressing parochial concerns would bring to the environmental agenda more expansive cultural and global concerns and thus multiply the interests involved in environmental policymaking.

The Confrontational Politics of Ecoterrorism

If the environmental justice movement is calling on environmentalism to shed its elitist character, expand its agenda to include urban and global issues, and in general demonstrate as much concern for human welfare as for natural resources, the so-called ecoterrorists are militantly pressing for just the opposite. Though known as much for their tactics as their mission, they represent an ideology more in tune with the historical conservationism of a John Muir than with any modern-day environmentalist and unabashedly place trees, wild animals, and sea mammals on equal footing with humans. As such, they represent a modest counterforce to the efforts exerted by environmental justice and other social and public health interests on the environmental establishment.

Ecoterrorism, sometimes known as "ecotage," is simply the direct action that individuals take to protect animals, trees, and other natural resources from what they regard as an outrageous assault by industrial and commercial society. Ecoterrorist actions include spiking trees destined for cutting, "decommissioning" construction equipment, lying down in front of bulldozers, "tree-sitting" in old-growth forests, burning developments that compromise the natural settings in which they are sited, throwing paint on fur coats, freeing animals from research laboratories, and sinking whaling ships. They call their disruptive tactics "monkeywrenching" and argue that the real terrorists are those who would destroy a redwood to make a picnic table. They depart from the essential character of the mainstream organizations from which many of their leaders came, both in their exaltation of nature and in their reliance on direct citizen action rather than lobbying to promote their cause. They decry the "corporatization" of the major environmental groups who, in their view, have become indistinguishable from their business counterparts and chide them for selling out their principles under the guise of reasonable compromise.

Despite their embrace of a lofty mission, ecoterrorists represent a radical fringe of the environmental movement. Although the most prominent of their groups—Earth First!—disdains violence, guns, and explosives, its members trade in what its founder, Dave Foreman, calls "guerrilla" tactics, "jiu-jitsu against the power structure." But even they have mellowed; they now engage principally in dramatic, headline-grabbing acts, such as taking up long-term residence in old-growth trees to forestall their "clearing." But its progeny, including groups like the Earth Liberation Front, who were behind the most costly single act of ecoterrorism—the 1998 burning of a development in Vail, Colorado—the Animal Liberation Front, and similar groups have been more destructive, if not more violent, in the service of "stopping the exploitation of the environment." One hesitates to call these groups organizations; rather they are a highly committed, loosely connected group of activists motivated by an old-style naturalism. Their low profile and the clandestine character of their network are no doubt attributable to their activities being illegal and prosecutable. Still, the consequences of these acts are significant. By 1998, no fewer than 1,500 acts of domestic ecoterrorism had been documented.

Ecoterrorist groups, of course, did not invent environmental activism. That is a legacy most recently of those who saw themselves as victims of the environmental degradation visited on their homes, schools, and workplaces. But the ecosaboteurs have brought the spirit of citizen action to the service not of the new environmental agenda, but of the old. In that they represent an interesting political anomaly. Their political effectiveness has been, and remains, marginal at best, however. Unlike the new breed of citizen activists, who seek to put new pressures on politicians to address the environmental conditions they feel threatened by, ecoterrorists seek to circumvent, rather than influence, institutional policymakers. It is, therefore, their dependence on intimidation rather than persuasion that makes them radical and, no doubt, keeps them from being anything other than an antagonist both to those in the mainstream who share some of their values, and even to many of their fellow coconspirators, who see them as undermining their own more tempered strategies. In a curious way, they have served the larger, long-term interests of the traditional environmental organizations, by making them look more moderate and rational, even as they remind those groups of what they see as the mainstream's abdication of their central mission.

Earlier in this chapter, and elsewhere in this book, I have spoken of what I term the gradual "domestication" of the environmental agenda, the evolution from "conservation" to "environmentalism," and the increasing attention paid to public health and urban concerns in a revised environmental agenda. I have identified as a turning point the publication of Rachel Carson's *Silent Spring*. Bringing public health into the public consciousness unwittingly unleashed a host of social justice concerns that had lain just below the surface. To be sure, the old, established environmental organizations have maintained their conservationist core, but others have been created, reinvented, and expanded at least in part to accommodate urban issues. Today, whether from an enlarged sense of public purpose or crass political expedience, a diverse group of environmental organizations, alliances, and coalitions have embraced any number of issues as important to inner-city residents as to their historic suburban constituency. But challenges such as those posed by the environmental justice movement demonstrate that the reconciliation still has a way to go. For our purposes, the continuing conflict is a quintessential demonstration that interests, not partisanship or ideology, drive environmental politics and, hence, policy.

CENTRAL IDEAS

National advocacy organizations carried the environmental agenda to Congress and the nation for most of the last century. But recent years have witnessed the appearance of a profusion of grassroots groups, who are concerned more with the conditions in their own communities than with the conservationist objectives of their forerunners. They have become a formidable political force and have goaded the government as well as mainstream organizations into addressing new issues in new ways. Less formidable, though newsworthy, have been the destructive activities of a band of ecoterrorists that, in ideology though not in practice, recall the philosophy of John Muir.

Chapter 8

The Greening of Business: Politics for Profits

Challenging existing environmental programs and priorities is not anti-environment. . . . The free market provides the basis for a superior approach to environmental policy.

The Competitive Enterprise Institute

For a long time there was a perception in industry that environmentalism was somehow a fringe activity, out of step with the mainstream of society. . . . Many manufacturing executives saw environmentalism as a nuisance and environmentalists as radicals. . . . But the most powerful environmentalist group in every modern society is now the general public.

E.S. Woolard

If the environmental advocacy groups have been the public's chief defenders against threats to the environment and the ecosystem, it is the business community whose activities have historically been seen as responsible for those threats. Industrial operations inevitably generate significant levels of air and water pollution and a steady stream of wastes, and building and infrastructure construction impose burdens on land and reduce its capacity to control flooding, drought, and soil and beach erosion. The products they provide for a consumer-oriented society exhaust natural resources faster than they can be replenished, when they can be replenished at all, and the energy supplies on which these activities depend impose their own risks to public health and safety, as accidents such as those at the Three Mile Island nuclear facility and the oil spill from the *Exxon Valdez* tanker at Prince William Sound attest. Because business operations pose such risks to the environment, they have been subject to increasingly stringent and comprehensive regulation by government. With the biggest stake in how environmental policy is framed, business has always been an aggressive and prominent participant in the policymaking process.

But even as businesses have continued to pursue their rational self-interest in the halls of Congress and with the administrative agencies, they have, over the past several decades, found a need to adopt new strategies and allies in an effort to respond to the public's growing environmental awareness. The result has been a complex, multifaceted effort to soften their image, break down the adversarial relationship between themselves and the environmentalists, and advance the notion that a thriving economy and a healthy environment are not mutually exclusive. Further, they have promoted the idea that environmentally responsible corporate behavior can make business as much a part of the solution as of the problem. Let's look at the evolution of business's posture over the past two decades or so and how it has played into the changing climate of environmental advocacy.

It should be noted at the outset that the business community represents a formidable force in policymaking. First, it has vastly greater financial resources than the environmental community, resources that have routinely been employed not only to support the campaigns of business-friendly candidates, but also to maintain a nationwide cadre of lobbyists pacing the halls of state and federal capitols, armed with voluminous information and detailed position papers on all issues that affect them. They have established national networks in the form of professional and trade associations that were originally designed to communicate with each other on the full range of common concerns but that have been directed in recent years to perform a lobbying function. The U.S. Chamber of Commerce, the National Association of Manufacturers, and the Small Business Association are but three of the largest and most influential alliances, but scores of other powerful groups representing individual industries, e.g., the Chemical Manufacturers Association, the American Petroleum Institute, the Association of American Railroads, the Motor Vehicle Manufacturers Association, the American Farm Bureau Federation, and the American Electric Power Service Corporation, all play significant lobbying roles in Washington as well as in state capitols.

Even more important than these organizational assets is a weapon that no environmental organization or any other pro-environmental group or interest can claim, their fundamental relationship to the economy generally and

to employment in particular. Historically, they have cultivated the notion in the public mind—with much success—that unreasonably regulating them will adversely affect the nation's economic well-being and cause job loss and dislocation. Thus, the big debate has historically been framed as "the economy versus the environment." It was the terms of that debate that they have had to change.

During the 1960s and the early 1970s, the political give-and-take between the environmentalists and business was not especially contentious. The post-World War II economic boom rendered the costs of meeting environmental concerns relatively innocuous, particularly since the EPA was having difficulty meeting the regulatory obligations imposed on it by an activist Congress. But the regulatory burdens on industrial and commercial activities began to take hold by the middle 1970s; the mainstream environmental organizations established a more dominant presence in Washington and, accordingly, took a more active and aggressive role in seeing that environmental standards were both sufficiently strong and enforced. By 1980, more than $55 billion was spent annually by the public and private sectors on pollution abatement, double that spent in 1972.

Business and the Reagan Agenda

These accelerating costs motivated the business community to match the environmentalists' increasingly active advocacy role in shaping environmental policy and practice. In particular, regulatory reform became their rallying cry, but they did not direct their efforts at the Congress. Rather, they focused on the executive branch and found a willing and powerful ally. Ronald Reagan, the avowedly probusiness president that they had waited for since the 1920s, was intent on curtailing what he saw as a government that was too large and intrusive and a body of environmental law that was wasteful and excessive. In his view, when government was needed, it should be that closest to the people; federal spending should be limited to essentials. In environmental law in particular, he found a quintessential example of bloated government, overregulation, resources squandered in the name of social theories, and free enterprise stifled.

President Reagan's effort to remedy the situation was supported from the outset by the National Republican Party platform of 1980 (which he, of course, had a role in shaping) that "declare[d] war on government regulation," and expressed the belief that "the marketplace, rather than bureaucrats, should regulate management decisions." Within weeks of taking office, he signed Executive Order no. 12291, which gave the Office of Management and Budget the authority and responsibility to review all rules for their *economic* consequences and which required that a cost-benefit analysis be conducted as a precondition to the adoption of new regulations to inform review of existing regulations and to be considered in proposed new legislation. The order also required administrative agencies to justify the need for the regulations and the consequences that they may have on those most affected by their implementation. With respect to potential harm, the probability of risk must outweigh the benefits of government noninterference in private activities. At the very least, the mandates of the order would have the effect of slowing down the rule-adoption process. A moratorium was imposed on many regulations to provide the opportunity to review them.

In addition to this "dream agenda," business had a virtual open door to the offices of the EPA under Administrator Anne Gorsuch Burford, who came with strong connections to the business community. The agency's rolls filled not only with former employees from regulated companies, but also industry representatives most affected by proposed regulations were admitted freely into decision-making proceedings to give "advice" during their development. Conversely, environmental organizations were generally excluded and public participation in environmental policy sessions virtually disappeared.

With respect to land use issues, Interior Secretary James Watt, who equated government regulation with socialism, was given a mandate to promote growth. Bringing strong western ties to the office, he supported an effort by a number of western states, dubbed the "Sagebrush Rebellion," that lobbied for a return of federal lands to local hands and the restoration of what it regarded as their lost property rights. Together they sought to open up public lands to mining, drilling, grazing, timber harvesting, and natural gas exploration. Extractive industries had high hopes of turning the corner on the "tyrannical" rule of the Bureau of Land Management and to gain a measure of parity with the federal government in managing their lands.

All this was, of course, music to the ears of business, who welcomed an administration that would promote their agenda vigorously and unapologetically. Economic factors were, once again, a major part of the mix, and the environmentalists were seemingly at bay. But, fortunately for the environmentalists, both Burford and Watt overplayed their hands, became entangled in scandal, and left the administration prematurely in disgrace. With them went any hope of fully realizing the agenda on President Reagan's desk as he took office, although the principal elements of that agenda—the decentralization of environmental power, the adoption of cost-benefit analyses as integral parts of rule-making, and the privatization of government functions—affect environmental policy to this day.

Although the scandals surrounding Anne Gorsuch Burford and James Watt set business's ambitions back to some extent, a series of major real-life events—most significantly the explosion at a chemical plant in Bhopal, India in 1984 and the massive oil spill in Alaska's Prince William Sound from the *Exxon Valdez* tanker in 1989 (with the distant echoes of the near disastrous nuclear accident at Three Mile Island of 1979 still in memory)—forced them to change direction more significantly. These events, of course, were given prominent media attention, reinvigorated the environmental consciousness of the public, and validated much of what the environmentalists had been saying about the risks posed by industrial activities. It apparently became clear to the business community that they had to focus more of their lobbying efforts on the public, who were understandably imploring their representatives to take stronger action to protect them from the risks to which they saw themselves increasingly subject. If big business were to head off a new round of even stricter regulations, public concerns clearly had to be addressed and negative attitudes toward them had to be changed.

Corporate Codes of Conduct

One of the initial elements of a new business strategy to accomplish this goal was to convince the public that business was as distressed by environmental accidents as the public, and that business would take every responsible measure to ensure a safe and clean environment. Thus, around the mid-1980s, several corporate interests began to adopt voluntary codes of environmental management practice. Though different in a number of particulars, all of these codes committed companies to adopt internal systems of environmentally conscious management and to track their progress toward goals they set for themselves at the outset. Three of these codes merit brief discussion.

The first of these codes, Responsible Care, was conceived by the Chemical Manufacturers Association (CMA), not surprisingly in the wake of the worst industrial accident in history, the explosion at a chemical plant in Bhopal, India in 1984, and a fortunately less serious one that followed a year later at a plant in Institute, West Virginia. These accidents tarnished the chemical industry's image and put communities hosting chemical facilities on edge. Recommended by CMA's Public Perception Committee, the openly admitted objective of Responsible Care was to improve their members' environmental, health, and safety performance to quiet critical public audiences. Responsible Care is a comprehensive program comprising a range of elements from community education and emergency response to pollution prevention, process safety, and employee health.

A second major code to be adopted was that put together by the Coalition for Environmentally Responsible Economies (CERES), formed to address environmental problems from the perspective of the marketplace. Its narrow purpose was to provide investors with environmental performance data that they could use along with other indices of performance in making their investment decisions. Its founder, Joan Bavaria, was insightful enough to realize that the availability of such data would foster greater trust between the member companies and the public. The core belief was that by paying attention to the quality of the environment, ecological and public health disasters such as the Bhopal accident and the *Exxon Valdez* spill that could threaten the economic viability of corporations could be avoided. The CERES coalition was more diverse than Responsible Care, comprising not only investors and investment professionals, but also environmental advocacy groups, labor unions, and even the State of California controller. The coalition's initial project was, tellingly, the Valdez Principles (though it was later renamed the CERES principles), a set of rules affirming the responsibility of corporations to conduct business in a way that is mindful of their stewardship of the environment. Specifically, the principles commit endorsers to protect the biosphere, use natural resources in a sustainable manner, reduce wastes and minimize their adverse impacts, use energy wisely, reduce risk, market only safe products, fully compensate for any damages for which they are responsible, fully disclose risks to employees and the public, appoint environmental managers or directors, and annually audit environmental performance. The CERES principles were not widely endorsed, but they did secure the participation of several major corporations, e.g., Sun Company, General Motors, and Polaroid when the principles were modified. The Valdez Principles got much publicity and surely raised corporate consciences even as they helped restore public confidence.

A year later, in 1990, the Global Environmental Management Initiative (GEMI) was born. It was the product of discussions among the environmental managers of chemical, pharmaceutical, electronic, and consumer product manufacturers who were motivated by concerns over potential environmental liability. They sought a forum to create a dialogue among themselves and with the public. It developed the Business Charter for Sustainable Development in conjunction with the Chamber of Commerce, which, like the other codes, invoked members to track environmental progress through self-audits, one result of which was the concept of "total quality environmental management." Through its national conferences, it publicizes its work to the public and to its sister business groups.

In sum, from their nature and timing, there seems little doubt that codes of conduct were undertaken in recognition of the need to regain public confidence. Hard-core environmentalists see them as just one of a number of business strategies to project an image of responsibility and concern for the environment without actually doing anything substantive. They look askance at the codes largely because their strictures are voluntary and because the public has only limited access to the performance data they track. Yet actions to which these codes commit their signers do go beyond the specific standards that would be required by regulation to the management of their internal systems. Also, in promoting the appointment of environmental managers to high-level positions in corporate organizational structures, they bring environmental issues into the boardrooms and the policy deliberations of big business.

Business Lobbying Through Surrogates

While corporations are directly taking on eroding public attitudes toward them by implementing these codes of conduct, their portfolios are being carried by two sets of coalitions or, perhaps more accurately, front groups. Though very different in strategy, posture, and character, the two share a common agenda that is favorable to, and endorsed and largely funded by, corporations, but act, at least superficially, independently. Together they constitute a genuine antienvironmental movement, or environmental "backlash," as it has been called. They remain major forces in environmental politics today.

This antienvironmental movement works and reworks a number of now-familiar themes:

- The economic costs that environmental regulation imposes on both the private and public sectors are crippling to the economy and taxpayers and disproportionate to the benefits it promotes.
- Natural resources are better protected and maintained by private ownership than by government regulation.
- Environmental regulations are not aimed at real, substantial health risks, but rather are keyed to the political agendas of environmental lobbyists and the budgetary needs and aspirations of governmental regulators.
- Virtually all of the allegedly catastrophic environmental threats— ozone depletion, overpopulation, acid rain, global climate change—are based on faulty, if not fraudulent, science.

- Environmentalists are technophobes or Luddites who deny or deplore the progress man has made and want to return to the precivilized, back-to-nature way of life of our earliest ancestors.
- Environmental concerns are most effectively and economically addressed by laissez-faire government and a free market economy.

Its adherents routinely assert not only that the sky hasn't fallen, but also that the environment, including our natural resources, is actually in better condition today than in years past, and getting better!

One of these front groups, called Wise Use, is a loose coalition of 400 or so small groups of ranchers, representatives of extractive industries, property rights activists, off-road vehicle recreationists, farmers, and right-wing ideologues. They coalesced in August 1988 at a national conference, the Multiple Use Strategy Conference, sponsored by the Center for the Defense of Free Enterprise, at which they set down a formal agenda. *Wise Use* is a term coined early in the twentieth century by Gifford Pinchot, the nation's first forester and chief of the U.S. Forest Service, to describe responsible conservation—the balancing of preservation and the profitable use of land in a manner that accomplishes both. But the modern antienvironmental alliance that uses the phrase for its name goes well beyond what Pinchot would have endorsed, advocating the virtually free and unfettered exploitation of land for people's needs.

Though Wise Use presents itself as a "down home, just plain folks" organization, it has enlisted the substantial support of corporations such as Chevron, Exxon, and the American Farm Bureau, as well as the expected help from national lobbies of the extractive industries. Picking up from the failures of James Watt and the Sagebrush Rebellion, Wise Use started out as a western property rights movement intent on repealing the Endangered Species Act, opposing repeal of the 1972 Mining Act, and generally opening up federal lands to development. But since the late 1980s, it has expanded to much of the country, challenging not only the wisdom but the very constitutional validity of government regulation of land. Wise Use has become the force that it has because it taps into popular sentiments—the sanctity of individual freedom and the evils of big government that compromise it. It is interesting to note that the movement rose to prominence at precisely the same time that a counterpart populist movement, called "environmental justice," was also developing into a national political force. In Wise Use, business has it own grassroots organization.

Different from Wise Use in many ways, but nevertheless serving business interests, is a group of conservative and libertarian think tanks that began taking an active role in shaping environmental policy in the 1980s, though most were established in the 1970s to address other economic and political issues. Whereas Wise Use is essentially western, these think tanks are headquartered principally in Washington, D.C. and while the focus of Wise Use is on land use and property rights, the focus of the think tanks is more often on the regulation of industrial and commercial activities. Bound together by the shared core conviction that the environment is best protected by the marketplace, not government, the promotion of free enterprise and limited government is their common mission.

Unlike Wise Use, the think tanks are funded principally by the corporate giants, though most frequently through their sponsored foundations rather than directly. This indirect method of funding and their careful avoidance of direct political action allow them to maintain tax-exempt status as educational institutions, despite the undeniable fact that they aggressively pursue a distinctly political agenda and influence both the public and lawmakers in a variety of ways.

The most influential of these think tanks, and the one most responsible for orchestrating the efforts of the group, is the Heritage Foundation, founded by Joseph Coors in 1973. It was, in fact, the Heritage Foundation that prepared Mandate for Leadership, a formal policy statement and agenda presented to Ronald Reagan as he entered office. Other major think tanks that often work in tandem with Heritage and other sister organizations, but always in sympathy with them, are the Cato Institute, founded in 1977; Consumer Alert, founded in 1977; the Reason Foundation, founded in 1978; the Political Economy Research Center, founded in 1980; the Competitive Enterprise Institute, founded in 1984; the Foundation for Research on Economics and the Environment, founded in 1986; and the Citizens for the Environment and the Science and Environmental Policy Project, both founded in 1990. Almost all of these are financed by the country's major corporate industrial groups—oil, pharmaceuticals, automobile manufacturing, tobacco, and chemicals; independent corporate giants such as Coca-Cola, Monsanto, IBM, Lockheed, and Alcoa; and other conservative foundations such as Scaife, Olin, Carthage, and Bradley.

These foundations promote their market-driven environmental policy prescription and, at the same time, undermine the efforts of advocates for governmental regulation of the environment though a number of diverse activities. Perhaps the most effective of their strategies is to literally flood the popular media—especially in editorials and letters to the editor columns—in rapid response to any legislative initiative, environmental event likely to generate a call for government action, or research finding alleging a new environmental threat or risk. It would be impossible to even calculate the number of stories, columns, letters, and editorials the group collectively has published in the last several years in newspapers and periodicals debunking global warming, for example, or chiding public fears of genetically modified food. Many of their own research reports become the subjects of newspaper "news" pieces. The Heritage Foundation, for example, once claimed to have gotten 200 media "hits" from one of its reports. The foundations also gain access to the public through radio and television talk shows, for which they make themselves freely available. The spokespeople for these think tanks have academic degrees, and their foundation affiliations seem impressive to citizens who are unaware of their sponsors.

The foundations also support the authorship and publication of popular science books that ridicule what they regard as the exaggerated claims of the environmentalists. Ronald Bailey's *Ecoscam: The False Prophets of Ecological Apocalypse*, sponsored by the Cato Institute; *The True State of the Planet*, an anthology of pieces challenging the environmental positions on everything from global warming, to pesticides, to biodiversity, a project of the Competitive Enterprise Institute; and *Environmental Gore: A Constructive Response to Earth in the Balance*, published by the Pacific Institute for Public Policy, are but three of a score of books readily available in bookstores throughout the country. As their titles suggest, these books chide the environmental community for fear-mongering through bad science and bad economics.

Finally, the foundations support their own legal foundations to challenge regulations for their constitutionality, consistency with legislative intent, or appropriateness of application. The Mountain States Legal Foundation, whose president formerly served James Watt in the Department of the Interior, is funded almost exclusively by a consortium of oil interests, and the Pacific Legal Foundation is financed by chemical manufacturers, power companies, real estate developers, and oil and timber companies. Even when these suits fail, they significantly delay the implementation of adopted regulations and divert government agency revenues from substantive work to legal fees, both of which serve their agenda, albeit indirectly.

Wise Use and corporate think tanks, then, both advance the business agenda with respect to environmental policy. They allow the business community to appear to stand apart from the political fray while their intermediaries work to shape public opinion on the issues most relevant to them. Meanwhile, the "big names" have been able to project an aura of responsibility in developing and implementing codes of conduct that ostensibly attest to their good environmental citizenship.

Back to Congress

The sluggish economy of the late 1980s and early 1990s made it difficult for the government to defend the strict environmental standards and rules adopted over the previous decade or two, and business's claims that they were responsible for job losses, plant relocations, and higher taxes were received with more sympathy. The climate was ripe for regulatory reform, and the so-called Republican revolution that swept into Congress in 1994 was prepared to provide it. Ironically, it presented the reverse situation from the one that prevailed in 1980; this time the Congress was pro-business and the president and vice president were sympathetic to environmental regulation. Business redirected at least some of its efforts once again toward Washington, but, as we shall see, it did not by any means abandon its direct appeals to the public.

The Republican congressional candidates had run on a broad program of reform that they called the Contract With America. Invisibly embedded in its many components were three environmental initiatives, all of which were unsuccessfully advanced by President Reagan: to establish a complicated procedure to analyze proposed regulations for their scientific soundness and cost-benefit ratios; to require compensation to property owners whose land uses were limited by environmental regulations and thus reduced in economic value; and to require the federal government to compensate states for all the costs of implementing new environmental programs, the so-called federal mandate-federal pay bill.

Each of these bills would have, in its own way, devastated environmental protection policies. And that was likely their intent. The business community saw in the contract's environmental bills another opportunity to enact at least a significant portion of the Reagan agenda. Although the measures were passed by the House of Representatives, only the "federal mandate-federal pay" bill passed the Senate and found its way into law, in amended form. The other measures failed, in all likelihood because then Senate President Robert Dole envisioned a run for the White House and couldn't brook what promised to be a bitter political battle.

Still, the very same business-friendly principles that President Reagan struggled to enact remained on the congressional agenda and continued to inform policy even if not enacted. Persistent conservative opposition to rigorous environmental regulation, supplemented by a deteriorating economy, began to take its toll on the EPA, and the much maligned and expensive "command and control" protocol of the 1970s and 1980s began to give way to a more conciliatory and cooperative relationship between business and government, as discussed in Chapter 3.

Business made gains with the EPA, but they came increasingly to the confusion that it was the public, not government, that would drive policy and that an even more comprehensive public relations effort was in order. The codes were developed out of the recognition that good environmental behavior is good public relations, but clearly more had to be done.

The 1990s ushered in a period of "corporate environmentalism." The clarion call was sounded by a cover story in the February 12, 1990, issue of *Fortune*, titled "The Environment: Business Joins the New Crusade." Identifying the environment as the "biggest business issue of the 1990s," it goes on

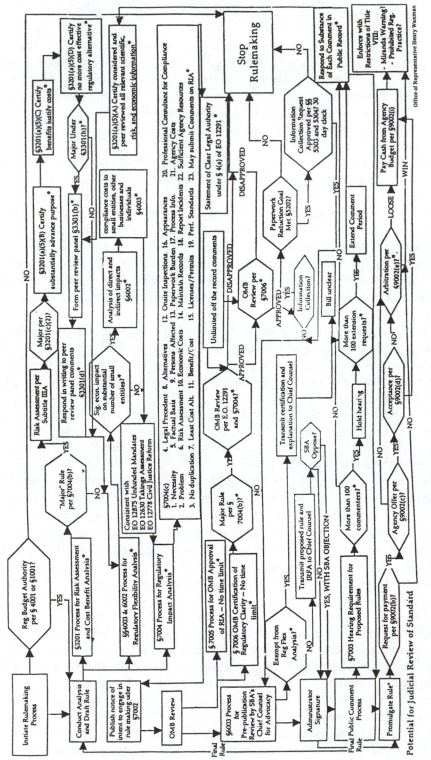

Mock up flow chart of HR9 of 1994, The Contract With America legislation on the regulation of toxic substances.

Office of Representative Henry Waxman

to detail the many initiatives that major corporate concerns were already undertaking to position themselves favorably with the public and capitalize on their environmental consciousness. During the next several years, daily newspapers, mass-circulation magazines, and industry publications alike were trumpeting business's newfound environmental stewardship and the economic, health, and social gains to be derived from the incorporation of environmental thinking into corporate planning. Even *Science*, the weekly publication of the American Association for the Advancement of Science, included an extensive, multipart section specifically devoted to "Environment and the Economy" in its June 25, 1993, issue.

That business and the environment are not only not incompatible but symbiotic was to become the core message of a national public relations blitz: "green" is practical *and* profitable. Business set out to send this message in a variety of ways. One way was to enter cooperative agreements with environmental advocacy groups. The one that got the most publicity was the 1992 accord fashioned between McDonald's and the Environmental Defense Fund (EDF) to phase out the Styrofoam clamshell packages in which hamburgers were served, but General Motors also signed an agreement with EDF to work together on a much broader range of issues: the scrapping of old vehicles, fuel efficiency standards, pollution-reduction credits, and urban smog, to name a few. Later, British Petroleum, one of the world's largest oil firms, undertook a joint project with EDF to address the emission of greenhouse gases as the source of global warming.

EDF was not the only mainstream environmental organization willing to work with, rather than against, big business. Chevron enlisted the World Wildlife Fund to advise it on environmental issues, and the Rainforest Action Network influenced Home Depot to phase out the sale of wood products from environmentally sensitive areas by 2002 and give preference to wood "certified" to have been the product of environmentally sound logging practices developed by a Forest Stewardship Council established in 1993. Three years later, Unilever, a major fish marketer, joined with the World Wildlife Fund to create a Marine Stewardship Council, which would address the dwindling world fish stocks by helping to ensure sustainable industry practices. On another front, California's Pacific Gas and Electric conducted an extensive, and expensive, study of energy efficiency.

These are only a few of the countless efforts entered into jointly by business concerns and environmental organizations. In reality, they serve the interests of both. Business has the money to fund these efforts and profits from association with environmental groups. For their part, the environmental groups secure the cooperation and assistance, to say nothing of funding, of the private sector, and at the same time dispel the perception of inflexibility of attitude and indifference to economics that dog them. Whatever their propaganda value, most of the initiatives accomplish meaningful purposes.

Of more questionable value was the explosion of "green products" that industry flooded the market with during the early 1990s, products touted as "biodegradable," "ozone friendly," "recyclable," "environmentally safe," "natural," "phosphate free," and even "reusable." Aimed at ecology-sensitive baby boomers, these "new" products multiplied by orders of magnitude during the early 1990s, and products sporting environment friendly labels reached nearly $9 billion in sales by mid-decade. But the trend trailed off as public skepticism about them grew and as consumer protection agencies and government watchdogs exposed many of their claims as meaningless at best, spurious at worst. Some that provide real though modest benefits, such as less and lighter packaging and more concentrated formulations, remain on store shelves and still attract environmentally conscious consumers, but clearly commercial marketers have retreated from this particular ploy to improve their image.

These products were but one of a number of promotional efforts conceived by a burgeoning sector of advertising specifically focused on maximizing their clients' economic benefit from a favorable environmental image. These agencies serve as consultants to companies, helping them target audiences, train their workforce, and "green" their operations. As business has grown to appreciate the policy impact that the general public exerts, they have devoted more of their time to attending the conferences, reading the newsletters, and studying the polls that these agencies sell them. In short, during the 1990s, the environment itself became an industry.

And though the influx of new "green" products has abated, environmental advertising has not. Even a cursory glance at general-circulation periodicals, for example, reveals ads posted by automobile manufacturers picturing their vehicles against mountainous crests or picturesque landscapes; by oil companies proudly publicizing programs they fund to demonstrate that the rare birds, turtles, or wildlife they depict can live in perfect harmony with a habitat that has been the site of recent "development"; or energy companies proudly announcing their plans for energy development in the post-fossil fuel era as their employees are pictured luxuriating in the sun-bathed natural settings. These are obvious attempts to reduce, or eliminate, the stigma these activities have carried, and they do so by association with favorable, natural elements. And the stream of editorial ads in the *New York Times* by any number of business interests publicizing their successful anti-pollution and energy efficiency programs continues unabated.

The environmentalists—at least the most uncompromising of them—have a name for the kind of image polishing that so-called green products, green advertising, and even some green projects trade in. They call it "greenwashing," new industry posturing to pursue the same old profit motive, a transparent attempt to pose as a friend of the environment without changing practices or procedures that would actually improve it. Harsh critiques of these practices have regularly appeared in a number of environmental journals, and Greenpeace, the most strident critic of all, has published an extensive report, "The Greenpeace Book of Greenwash," excoriating such practices principally by transnational corporations.

Still, the appeals of business toward the public have had some effect. Americans are pulled both by their genuine desire for a safe, clean, and healthy environment and by an almost equal desire for a strong economy, full employment, and low prices. In the picture that business now paints, the public can have both, which is why they have been as receptive as they have to these appeals. Even though, at least superficially, business has turned away from traditional politics and toward the politics of populism, environmental policy is still very much in their sights. At the outset of the George W. Bush administration, a major battle over oil drilling in the Arctic National Wildlife Refuge is about to break out. Pro-business congressional members, as well as the White House, are convinced that oil development in that environmentally sensitive region can be conducted harmlessly and compatibly, with the native peoples, wildlife, and other living things that inhabit it. A recent oil company ad picturing waterbirds floating peacefully along against a bucolic backdrop is, almost surely, a proactive effort to prepare the public for that major policy battle.

Before closing the discussion of business as an environmental interest group, it is appropriate to take at least a brief look at a subset of the business community, organized labor. Labor has kept a relatively low environmental profile, most often spending its political capital on issues that more directly affect workers' daily lives. The roles it has played over the years in environmental policy debates have been somewhat mixed, for no sector has as much reason to play it both ways as it does. Its interests clearly overlap those of employers when jobs are threatened and the economy is flagging. This is why, during the 1970s, labor took the side of management against the environmentalists when major industrial projects such as the supersonic transport (SST) or the Alaska pipeline were debated or when continuing construction of nuclear power plants was threatened, because these promised to generate jobs and help power the economy.

On the other hand, throughout most of the postwar years, when the economy was humming along and workers were enjoying steadily increasing wages and improving working conditions, labor had the luxury of thinking about quality-of-life issues and participated actively in congressional deliberations on matters such as clean air and water and even wilderness protection. Despite its support of major industrial projects during the 1970s noted above, labor's principal posture was strongly pro-environmental. The AFL/CIO celebrated the first Earth Day in the June 1970 issue of the *Federationist*, the organization's official monthly magazine. Its lead editorial and stories lamented the nation's environmental deterioration, affirmed the need to readjust the balance between the environment and the economy, and carved out the role that labor should play in solving problems like overpopulation. Almost a decade later, its October 1979 issue renews labor's commitment to a clean environment and dubbed as "environmental blackmail" the notion that environmentalism is contrary to a good economy, documenting a minor increase in cost of living attributable to environmental measures but a significant increase in the jobs they created. The same view was repeated during the decade, most notably by Leonard Woodcock, President of the United Auto Workers, in a 1976 speech. But in the 1980s, labor took aggressive action when the two competing forces acting on it came together in proposed "worker right to know" legislation. Passage of this legislation would have adversely affected business's bottom line, and, by extension, employees' economic security. But it would have forced management to disclose otherwise unknown or undetectable workplace hazards to which workers were routinely exposed. Thus, many union leaders saw in labor's pivotal position on this controversial legislation leverage in negotiating contracts and tried to extract whatever concessions they could as a bargaining chip for their support.

Despite their early history of pro-environmental activity, which was an integral part of a broad social agenda to which labor leaders committed their members, labor has recently been associated in the public mind more with business, largely because strong environmental regulation is perceived, rightly or wrongly, to undermine the economy. The gradual erosion of union membership and the corresponding reduction of its influence over the past decade or two have only reinforced that disposition.

It must be remembered that labor has stood apart from business on specific environmental issues when it was in labor's interest to do so; labor is by no means in the pocket of corporate managers. Moreover, the disparity between the interests of business and labor was also dramatized by the pitched battle over the North American Free Trade Agreement (NAFTA) in the mid-1990s. Labor saw the treaty as a direct threat to their jobs and fought it bitterly, against the strong support of big business. The supreme irony is that the lower production costs and price of foreign products—attributable among other factors to laxer environmental standards abroad—are what make free trade attractive to U.S. business and threatening to labor. That subject is discussed in more detail in Chapter 9.

CENTRAL IDEAS

As the principal target of environmental regulation, business historically exercised its political muscle in Congress and with the administrative agencies. But as power gradually shifted away from government and environmentalism became a popular cause, business has had to redirect its lobbying efforts, adopting corporate codes of conduct, funding front groups, and conducting a systematic advertising campaign highlighting "green" products and "green" imagery. They have also had to work jointly with, and often fund, environmental advocacy groups. Over the past half century, organized labor has, by turns, sided with business and with the environmental community as the economy and national and world events have dictated.

Chapter 9

Global Pressures and Domestic Environmental Politics

The WTO has been granted spectacular powers to challenge every nation's environmental laws. So far, its victims include dolphins, sea turtles, clean water, clean air, safe food, family farms, and democracy itself. But it's just getting started.

Turning Point Project

World trade is an odd choice as the root of Bad Things. Trade is about as neutral a phenomenon as one can imagine. To blame trade for a long list of injustices makes no more sense than to blame the alphabet for objectionable ideas.

Bernard Wasow

If, as previous chapters argue, business interests have had to make significant concessions to the environmental and social demands of a changing society, they seem to be getting new life as rapacious globalism blurs the distinct character of laws and values of sovereign nations. Today, many environmental gains achieved over the past decades in the U.S. seem to be up for review as they increasingly affect other nations. Some environmentalists have even complained that the guts of the past 20 years of environmental policy are being eroded. That may be too pessimistic an evaluation, but it is clear that as we move to a world stage, some policies are being revisited as the interests of other nations must be accommodated.

The confluence of environmental, business, and labor interests that weighed in as the nation deliberated over the North American Free Trade Agreement (NAFTA) foreshadowed what have become some of the most far-reaching and contentious environmental policy issues of the decade, as well as the entrance onto the political scene of new pressures, new authorities, and new rules from abroad. At first glance, it may seem inconsistent to round out the chapters on interest groups with one about the effect of international pressure on domestic environmental politics, since this book has argued that the interests served by U.S. environmentalism have become increasingly local, even parochial. Yet, in a complex way, the inevitable move toward globalization logically extends issues and trends discussed throughout this book, juggles old alliances and forges new ones, and brings old environmental issues to new venues. The way these factors play out will affect environmental policy here as well as internationally for the foreseeable future.

The 1990 Earth Day invocation to "Think globally, act locally" was probably the first nationally publicized announcement that we are entering a new world. Though the U.S. is a signatory to more than 150 international treaties dealing with the environment over the past century, the public has been largely unaware of them, and they have not substantially affected our lives or our nation's laws. But during the 1990s, global issues have come to the fore, and actively engaged a wide range of business, environmental, religious, and social justice groups as well as the general public. Today, as the move toward globalization is accelerating at a headlong pace, the impact of the policies and practices of one country on others is increasing proportionately. New international authorities are needed to establish rules to cushion these impacts and new treaties will commit signatories to observe them. As these policies are being debated and instituted, however, national interests and values are at stake. In no area is state sovereignty more threatened than in the area of environmental protection. The consequences of the enormous political and economic influence of the U.S. on the world have been thoroughly examined and analyzed by scholars around the globe, but the impact of certain of the world's conditions and forces on our nation's environmental laws also merit at least brief attention.

The most publicized international environmental issue in recent years has been global climate change, the phenomenon of the world's gradual warming due to the buildup of so-called greenhouse gases, principal among which is carbon dioxide. Global climate change, or global warming as it is popularly called, has been kept alive as an issue in this country by virtually all mainstream environmental organizations, which regard it as the quintessential challenge facing the planet, and by the U.S. and European scientific communities. For reasons discussed throughout this book, however, it has never been regarded by Americans as a defining issue. It doesn't affect their lives immediately or directly. Their indifference has been cultivated by a cadre of "global skeptics" who continually chide environmentalists here and abroad for their reliance on what the skeptics regard as highly speculative science.

The U.S. public's attitude toward global climate change is not shared by the rest of the world, whose scientists continually study the phenomenon and continue to find increasing evidence of is validity. World concern culminated in a treaty signed in Kyoto, Japan in December 1997, to which scores of nations, among them the U.S., committed to the reduction of the carbon dioxide gases alleged to be at the core of the problem. Despite the formal action taken by the U.S. representatives, spurred by Vice President Al Gore's personal appeal to them at the meeting, only scattered voluntary acts by individual corporations have been taken to meet our goals. Americans are generally disposed to honor the commitments made to the world community at Kyoto, but business has successfully convinced the public that meeting those obligations will result in higher energy costs and favors deferring any action unless and until other major developing countries meet their corresponding obligations. The only direct congressional action on the issue, Senate Resolution 98 of 1997, conditioning U.S. participation on a proportionate response from other nations, was passed by a 95–0 vote, indicative of the political weakness of the pro-environment position. Legislative riders continually defeat bills appropriating monies to address in a piecemeal fashion one or another aspect of the problem. More recently, the George W. Bush administration, in a bow to business, has backtracked on CO_2 emissions from utilities, and effectively canceled U.S. participation in the Kyoto Treaty.

A much more publicly charged issue, and one that has surfaced only in the last few years, has developed from the application of biotechnology to food production. This practice was initiated in the mid-1990s by multinational biotechnology corporations and U.S. farmers without public notice, and significant percentages of some affected crops, most notably corn and soybeans, were in the market before consumers were aware that they were eating them. It wasn't U.S. citizens who were most concerned at first but Europeans, especially the English. No doubt disaffected in some measure by their nightmarish experience with mad cow disease, the British reacted with sharp disfavor to genetically altered foods, and the citizens of many other European countries were similarly outraged. Within a relatively short period, genetically modified organisms (GMOs) were the subject of actions ranging from outright bans to calls that genetically altered foods be distinctly labeled as such so that consumers could avoid them if they so desired. Belatedly, American consumers have become increasingly uneasy as well, no doubt partly as a result of the European reaction. To date, however, no formal action has been taken by U.S. officials, though the U.S. Department of Agriculture and the EPA are continually conducting studies and hearings.

In this area and in others involving world trade, international opinion and policy are having increasing influence on U.S. environmental policy. Shortly after the disclosure of the nature and extent of the application of biotechnology to food production, the European Union (EU) began to establish rules governing the regulation of GMOs. Although U.S. authorities quickly approve genetically modified foods for market and export on the grounds that they are substantially equivalent to their nongenetic counterparts, the EU approval process is much more protracted, and labeling of GMO food products, which U.S. farmers and biotechnology companies regard as tantamount to a stigmatization, is required. The international reaction has had a significant effect on the exportation and sale of GMO foods abroad, a substantial percentage of the American agricultural market, and on GMO management here. Many farmers have cut back on their use of GMOs, and others have begun to undertake the costly process of segregating GMO and non-GMO foodstuffs. Other food technologies, e.g., beef hormones and bovine growth hormones, are routinely applied in the U.S., but both are restricted by the EU. Eco-labeling of packaging, recycling, and production processes is common in Europe but relatively rarer in the U.S.

Aggressive environmental policies such as these by the European Union can be viewed in a number of ways. American interests see them as protectionist; ways to secure a market advantage for their own products. Europeans, conversely, see trade policies as a way to achieve environmental goals. Other EU policies (e.g. requiring manufacturers to take back and recycle automobiles and electronics, limiting flexibility measures that countries can take to meet their greenhouse gas emissions quotas to 50%) similarly promote environmental protection at the expense of businesses. Whatever its motivation, the European Union has become a force in promoting global environmental protection, and American business interests are accordingly disadvantaged by their rules.

The U.S., as well as other nations, certainly can challenge strictures it regards as protectionist through the World Trade Organization (WTO) review process. The WTO, created in January 1995 to oversee international trade, settle disputes between nations, and organize trade negotiations, represents more than 130 members and thus carries enormous political power. In a global economic climate in which products, raw materials, and even services routinely cross national boundaries, it performs an essential function. In single-mindedly promoting and regulating free trade, however, it has found itself in the middle of a political storm, because promoting free trade, as the WTO conceives it, frequently sacrifices environmental standards, as well as worker rights.

Even in its short history, the WTO has forced the U.S. to alter some of its laws and regulations. For example, the WTO ruled that U.S. regulations under the Clean Air Act, which established high standards for gasoline, unfairly disadvantaged foreign oil companies, forcing the U.S. to rewrite its regulations. In an effort to protect sea turtles, the U.S., under authority of the Endangered Species Act, imposed a ban on shrimp imports from countries whose boats used nets not equipped with turtle excluder devices to protect these rare marine reptiles from suffocating. Shrimp nets kill up to 150,000 sea turtles per year. The WTO again ruled the ban an unjustifiable burden on free trade because it was not imposed uniformly on all countries. Similarly, in 1991, a provision of the Marine Mammal Protection Act prohibiting the import of tuna from countries whose fishermen were killing dolphins with their nets had been found to be inconsistent with the General Agreement on Tariffs and Trade, predecessor to the WTO.

Thus, in its pursuit of free trade, the WTO has effectively repealed or amended environmental laws and regulations democratically enacted over the past 25 years. It has done so consistent with several guiding principles. First, the method by which a product is made cannot be the basis of a discriminatory rule or action. This was the underlying principle in its decision regarding the tuna nets that killed dolphins; *how* the tuna were caught could not be regulated. If brought before the WTO, however, the EU's position on genetically modified foods could be undermined by this principle. Similarly, it could be undone by the WTO's belief that labels themselves are an undue restraint of trade.

Another of the WTO's guiding principles is that restrictions cannot be more obstructive to trade than deemed necessary by a broad consensus of the world community. This limitation would almost invariably eliminate bans of pollutants or practices to achieve environmental goals, because bans can rarely surmount this hurdle. Perhaps more significant, it would seemingly conflict with the increasing application of the Precautionary Principle discussed in Chapter 5. In fact, Principle 15 of the Rio Conference on the Environment and Development Declaration states that "the precautionary principle shall be widely applied.... Lack of scientific certainty shall not be a reason to postpone cost-effective measures to prevent threats of serious environmental damage." Environmentalists regard the present WTO principle conditioning restrictions on consensus as a "race to the bottom" for environmental protection measures. Requiring the concurrence of the community of nations to salvage environmental measures means that it is the lowest common denominator that will survive as a standard.

But the principal goal of the WTO, representing the economic aspirations of its member nations and the world's most powerful corporations, is free and open trade, not the prevention of environmental damage. If the WTO sees a nation's environmental law as a barrier to trade, that law can be abrogated and sanctions recommended. It has little of the kind of opposition that prevails in a typical democratic legislative context, nor can it. First, free trade has been embraced almost universally as an unqualified "good." Second, as a practical matter, disputes between nations are initially heard by a panel of three trade experts; appeals go to a larger body. Both, however, are conducted behind closed doors. Even the briefs filed by the contending parties are not shared with the public by the WTO; only the parties themselves can do so. Thus, neither environmental expertise nor public opinion can inform their deliberations.

The European Union and the World Trade Organization, then, have become de facto environmental interest groups that are playing increasingly significant roles in U.S. environmental politics. The EU, wittingly or unwittingly, is using trade restrictions to achieve environmental ends, forcing the U.S. to balance pressures from corporate citizens to maximize profits with the public's demand for a safer and cleaner environment. The mandate of the WTO to liberalize trade, however, is undermining many of the hard-won environmental practices put in place over the past several decades—practices such as pollution prevention, citizen participation, and the public's right to know.

If, as Chapter 8 documented, business interests in the U.S. have had to accommodate many of their practices to the demands of the environmental community and to the expectations of an environmentally sensitive public, in the international forum the situation is virtually the reverse. In this venue, closed deliberations, rather than openly democratic institutions make policy. Furthermore, environmental regulations must be justified as necessary by a broad consensus of those potentially affected, rather than by the one most appropriate to the environmental or public health problems to be addressed. The international forum favors the least restrictive measure consistent with free trade, rather than a more cautious "better to be safe than sorry" principle. Also, problems are to be dealt with reactively, when they manifest themselves, rather than proactively, at their sources. In short, in the name of free trade, established environmental policies are being challenged and accordingly modified or abandoned.

These practices and principles are in fact being publicly challenged. In Seattle, in December 1999, a consortium of environmental, labor, religious, and human rights groups and other ad hoc coalitions staged a massive street protest against the WTO. They succeeded in stifling its proceedings. The protesters had a long list of grievances against the WTO that they wanted to air, but their principal mission was to force it to open its decision making process to the public. Although the protesters did not specifically accomplish their goals—such protests rarely do—they did secure a significant measure of publicity for their causes. Characteristically, the very media that have largely ignored the actions of the WTO in the years since its creation—even those that undermined the democratic process—did cover the protest. Unlike the decisions of a foreign trade body, the protest itself was newsworthy, but apparently its issues were not.

As evident in the composition of the protesters, the backlash against the WTO has brought together a variety of disparate groups and interests that superficially have little in common but resemble a paradigm of the environmental justice movement in its grassroots, multidimensional constituency. The multitude of groups suggests that it is not solely environmental protection that is being sacrificed to trade. Globalization has, most important, reunited environmentalists and labor, who need each other politically. After all, high wages and quality working conditions can also be deemed obstructive to free trade. U.S. factories have, in fact, gravitated to foreign countries, where both environmental and labor standards are lower, thus making the prices of their products more competitive but threatening the loss of American jobs.

The causes behind the protests in Seattle—and, six months later, in Washington, D.C. by a coalition that calls itself the Mobilization for Global Justice against the policies of the World Bank and the International Monetary Fund with respect to third-world debt relief—have penetrated electoral politics. At the WTO meeting in Seattle, President Clinton called on the WTO to put the environment on the agenda and to make its meetings more public. During the 2000 election campaign, Vice President Gore, in response to concerns from environmentalists that free-trade policies threaten natural resources, proposed a new global environmental organization to participate in cross-border dispute resolution.

More substantively, in July 2000, a group of 50 transnational corporations signed an agreement with a broad coalition of watchdog groups to voluntarily protect workers and the environment. The agreement is to be implemented through a series of partnerships among governments, businesses, and nongovernmental organizations. Whether this agreement is substantive or merely a public relations effort remains to be seen, but it does represent a tacit recognition on the part of big business that the environmental, labor, and social justice issues raised by both their domestic and global operations need to be addressed.

In their uphill battle, the environmentalists have been given some unexpected reinforcements. On Earth Day 1997, the Department of State issued what must be regarded as an extraordinary document: *Environmental Diplomacy: The Environment and U.S. Foreign Policy*. In this report, Madeleine Albright, then Secretary of State, identified five major environmental problems that the world faces and that the U.S. is committed to help resolve. What is extraordinary, especially for an official government document, is that Secretary Albright tied the solution of these problems to national security. Successfully addressing these problems—climate change, toxic chemicals, species extinction, deforestation, and marine degradation—the secretary affirmed, not only will improve the quality of life of the world's population but also will reduce one of the sources of human conflict: the competition for scarce natural resources. In making the resolution of environmental problems an element of foreign policy, the Department of State is tying nothing less than world peace to the satisfactory resolution of environmental problems.

Domestic and international environmental conditions are politically intertwined, and U.S. policies will henceforth be affected by global factors and actors. As U.S. environmentalists begin to compete on a world stage, they may have a new and formidable, though unexpected, ally—the U.S. Department of State if the Bush administration maintains this initiative. As we shall see in Chapter 10, a new weapon in their arsenal extends their reach to every part of the globe—the Internet, the ultimate informational and organizational tool.

CENTRAL IDEAS

As globalization proceeds unabated, new international factors are influencing U.S. environmental policy. Prominent among these is the European Union, which has established stronger environmental standards than American business interests are comfortable with. More significantly, the WTO, in the service of free trade worldwide, is effectively repealing democratically enacted U.S. laws and undermining policy gains secured over the past 25 years. Its operations and practices are being challenged, however, and the State Department has tied solution of environmental problems to U.S. and world security.

Chapter 10

On-Line Activism and the New Environmental Politics

So while some of us may dread the prospective democratization of democracy via the Internet—and some may cringe at the thought that an on-line campaign by core supporters helped lift Jesse Ventura off the mat and into the Minnesota governor's chair—make no mistake: The full interactive potential of the Internet offers a real chance to restore some purpose to our politics by restoring some power to our people.

Doug Bailey

As recently as 1996 . . . most activists had no idea of the Internet's value as an advocacy tool. Now, every advocacy group worthy of the name is working on-line. . . . Its power rests with being able to mobilize supporters to take political action. . . . [it] seems to be a natural weapon for those who are in the business of "playing offense." While it can be used in a defensive mode, the Net is much better at starting a battle than at ending one.

Tom Price

Previous chapters documented two seemingly antithetical trends in the evolution of environmental politics in recent years. One shows the broad-based national environmental organizations grudgingly giving way to smaller, grassroots groups focused on local issues and problems. The other documents the growing influence on our environmental politics of conditions and concerns in third-world countries, the consequences of globalization, and the demands and restrictions imposed by international organizations and treaties. Without the concurrent development of a third element, information and telecommunications technology in general and the Internet in particular, however, the realignment of interests that those trends represent would not have taken on the shape and force that they have, and U.S. environmental politics would be very different today.

It goes without saying that the technological revolution in communications, in the space of a decade or two, has made an impact on almost every aspect of our lives and transformed our society and culture in ways that are still being calculated and evaluated. Its effect on environmental politics, however, even to this point, has been so far-reaching and profound that it has turned it on its head. Whereas, as argued in Chapters 2 and 3, the nation's environmental agenda and policies were shaped by formidable interest groups with access to and a large degree of control over the lawmaking process, the Internet has leveled the playing field, democratized the process, fueled a burgeoning populist agenda, and caused national governments and even transnational corporations in some cases to run for cover.

Let's look first at how this new technology has informed and educated citizens, both with respect to the policymaking process and on a whole range of issues. We then take up its effect as an organizational tool for individuals, government organizations and agencies, and independent groups. Finally we look at how, in providing these services, technology has fostered the transformations discussed in Chapters 5 and 8.

This kind of information may seem relatively basic and of marginal utility, and, to the extent that most of it is available from other sources, it is. Securing it, however, has always been complicated and time-consuming and required more time and effort than most people were willing to give, which is why so many citizens have historically sat on the political sidelines and allowed the organizations and government agencies to represent their environmental interests. As more and more information is put on-line, however, and as growing numbers of citizens are connecting to the Net, individual and small-group observation of and involvement in policymaking have increased rapidly.

It should be pointed out that even this information-seeking process is, to a degree, interactive. Members of Congress have their own websites on which they identify the bills and issues most important to them, promote initiatives they intend to pursue, and, in some cases, justify their votes on high-profile legislation. The president does the same, using the White House website to highlight his agenda, reproduce texts of his speeches, and furnish copies of press releases, among many other communications. Thus, an Internet user seeking simply to secure some basic legislative information is commonly invited into a kind of dialogue with the source of that information. In this political climate, no one, it seems, can pass up the opportunity to get his or her message out.

The World Wide Web as Content Provider

Not only does the Internet facilitate citizen participation, but also participants have never been more conversant with the relevant background and research. Today, hundreds, probably thousands, of sites are crammed with information on major subjects and topic areas. There is, in fact, no environmental problem or concern for which there isn't a vast mine of information, analyzed from every legal, fiscal, political, and policy perspective. Countless institutes, foundations, and think tanks have found a means to disseminate their reports, analyses, and other publications to the public in an efficient, economical, and paperless way. Environmental organizations devote portions of their websites to issues central to their mission, although, unsurprisingly, they take on an advocacy character. In addition, an incalculable number of websites of smaller organizations, alliances, coalitions, and interests supplement the mainstream menus, most of them restricted to a narrower agenda and sometimes to a single issue. Many of these are publicized and accessed through the network of links from other sources.

The inquiring public also has available a large and increasing number of databases. Much, perhaps most, of that information is provided by government and is rendered, for example on EPA's "Envirofacts Warehouse" site, not merely in the aggregate but also broken down by geographical area, facility, population, company, or pollutant. Indeed, almost as soon as any information that industry is required to report is compiled, it is made readily accessible to any citizen in the world without cost. "Envirofacts" is but one of a number of websites maintained by the EPA that provide a broad spectrum of data across a number of vectors. Surf Your Watershed, as its name suggests, identifies chemical pollutants in local watersheds. The Sector Facilities Indexing Project tracks the environmental performance of facilities in major industries. Environmental Monitoring for Public Access and Community Tracking covers public health and environmental conditions in local communities, and the Integrated Risk Information System reports on human health effects from exposure to different substances. Right to Know Data, Sector Facility Indexing, and the Toxic Release Inventory of hazardous emissions, which must be made public under the Superfund law, further exemplify the range and specificity of collected, collated, and distributed information.

Of course, governments at all levels must take care not to interpret the data beyond what can be definitively proved. Thus, they deal primarily in raw numbers. Not so with environmental organizations. Although several mainstream organizations provide their own scientific databases, one maintained by Environmental Defense (formerly the Environmental Defense Fund) and called the Scorecard probably has generated the most attention and the most chagrin in the private sector. It claims to provide information about 6,800 chemicals released by manufacturing facilities, but it also compares and evaluates pollution conditions in specific geographical areas across the country. It ranks, for example, the cancer risk to residents in the top 100 zip codes. In the course of providing visitors with information on the air quality, water quality, land contamination, agricultural pollution, and other adverse environmental conditions prevailing in their localities, Environmental Defense is not reticent to qualitatively characterize the hazards as well as enumerate them. This site is, then, generally more interpretive than the federal government sites, although most of the information it transmits is gleaned from public sources.

Of course, not all of the databases are concerned with chemicals and hazards. Some are more political. The Environmental Working Group, for one,

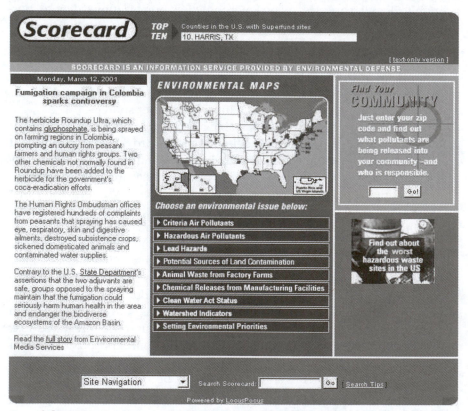

About Your Community

from the Community section of Scorecard.

Your Zip Code: 78701   Your Community: TRAVIS County

Welcome to Scorecard!
If you'd like to personalize Scorecard so that it remembers your community, you may register here.

WATER

Clean Water Act Status: Do Waterbodies in Your Community Meet Clean Water Act Standards?

- 8 % of surface waters in TRAVIS County have beneficial uses which are impaired or threatened. (Reports may be incomplete)
  - Some Rivers, Streams and Creeks are impaired by Pathogens and Salinity/TDS/Chlorides
  - Some Lakes, Reservoirs and Ponds are impaired by Pathogens and Pesticides
- The leading sources of water quality problems are Nonpoint Sources and Municipal Point Sources
- Learn more about Clean Water Act compliance in your community

Watershed Indicators: How Healthy Are Your Watersheds?

- TRAVIS County contains a portion of 5 watersheds.
  - EPA has determined that 2 have less serious water quality problems
- Learn more about watershed health in your community

AIR

Air Quality: Health Risks from Hazardous Air Pollutants

- In 1990, this county ranked among the worst 20% of all counties in the US in terms of noncancer hazards from hazardous air pollutants
  - 546,135 people in TRAVIS County face a cancer risk more than 100 times the goal set by the Clean Air Act.
    - 77% of the air cancer risk is from mobile sources
    - 23% of the air cancer risk is from area sources
    - 0.36% of the air cancer risk is from point sources
  - What's Your Risk?
  - Learn more about hazardous air pollutants in your community

Air Quality: Does Your Community Meet Clean Air Act Standards?

- In 1996, this county ranked among the worst 20% of all counties in the US in terms of emissions of nitrogen oxides (ozone season daily average)
- How Clean is Your Air?
- Who is Polluting Your Air with nitrogen oxides?
- Learn more about criteria air pollutants in your community

LAND

Potential Sources of Land Contamination

- No waste sites in TRAVIS County are on EPA's National Priority List of Superfund sites.

Lead Hazards

- 3400 houses in TRAVIS County have a high risk of lead hazards.

WASTE

Toxic Chemical Releases from Manufacturing Facilities

- In 1998, this county ranked in the top 20% of all counties in the US in terms of total off-site transfers
- Who is Polluting Your Community?
- What are the Major Pollutants?
- Learn more about pollution from manufacturing plants in your community

Agricultural Pollution

- How does Your Community Compare?
- Learn more about animal waste from factory farms in your community

SETTING ENVIRONMENTAL PRIORITIES

- One of the top-ranked environmental problems in your state is Stratospheric ozone depletion
- Learn more about environmental priorities in your state

Explore the Maps: See how air pollution in your area compares with other communities. Locate polluters, and see how close they are to your home or workplace.

Compare This Community to Others

Take Action: Send faxes to the top-ranked polluters in your area, send email to government officials, volunteer with environmental organizations in your area, or join Scorecard's online community forum.

For information about another community, enter the zip code: [   ] Go

---

Benjamin_Smith@environmentaldefense.org

ENVIRONMENTAL WORKING GROUP

Dirty Money: Tracking the PACs

Dirty Money Home

The Dirty PACs

Top Dirty Money Takers

Frequently Asked Questions

Search by Candidate
[Last name] Go

Search by PAC
[PAC name] Go

Data updated:
1/10/2001

Every year, polluting industries and their lobbyists donate millions of dollars to politicians' re-election campaigns. As a result, polluters have far more access to political decision makers -- and far more influence over environmental decisions -- than ordinary citizens do.

EWG's **Dirty Money Tracker**, based on data collected by the Federal Election Commission, lets you see for yourself how money from anti-environmental corporations and coalitions affects environmental decision makers.

Search by Candidate

Type in the last name of a candidate:

[   ]

[Find Candidate]

Search by PAC

Type in the name of a company:

[   ]

[Find Company]

How Does Your State Compare?

See how your state's politicians rank.

[Choose Your State ▾]
[Find State]

The Big Winners

The **top 10 recipients** for each of our five categories of polluters.

--OR--

Search by Type of Polluter

The candidates who took the most money from industry coalitions with shared anti-environmental interests.

[Choose A Category ▾]
[Find Top Recipients]

Note to Macintosh users: This site is best viewed with Netscape. (download it for free here)

Environmental Working Group • 1718 Connecticut Ave., N.W., Suite 600
Washington, DC 20009 • info@ewg.org

Reprinted with the permission of the Environmental Working Group, from www.ewg.org

regards itself as an environmental content provider for public interest groups and concerned citizens whose mission is to protect the environment. It prepares reports and articles on substantive subjects, but much of its most effective work lies in its electronic resources. It has searchable databases on subjects ranging from farm subsidies to violations of federal standards for drinking water to Clean Water Act permit backlogs. Perhaps its highest profile work is in the area of pure politics, in particular helping peo-

ENVIRONMENTAL WORKING GROUP

The Big Winners in 2000

The Top Campaign Contribution Recipients of Dirty Money from Various Anti-Environmental Interest Groups

Some politicians hit the jackpot by tapping polluter's dirty money. The following are the top 10 incumbent recipients from each general special interest group based on their PAC contributions for this election cycle.

- Click on "more" for the top recipients in that category by coalition.
- See a historical record of the top recipients for the last 3 election cycles.

Sidebar navigation: Dirty Money Home · The Dirty PACs · Top Dirty Money Takers · Frequently Asked Questions · Search by Candidate [Last name] [Go] · Search by PAC [PAC name] [Go] · Data updated: 1/10/2001

Rank	Dirty Air (more)		Dirty Water (more)		Global Warming (more)		Public Land Use (more)		Toxics (more)		Pesticides (more)	
1	Spencer Abraham (R-MI)	$549,663	Spencer Abraham (R-MI)	$206,438	Spencer Abraham (R-MI)	$544,911	Slade Gorton (R-WA)	$203,483	Spencer Abraham (R-MI)	$543,324	Charles W Stenholm (D-TX-17)	$74,290
2	Richard J Santorum (R-PA)	$449,744	Richard J Santorum (R-PA)	$197,412	Spencer Abraham (R-MI)	$446,634	Spencer Abraham (R-MI)	$201,216	Rick A Lazio (R-NY-2)	$456,708	Spencer Abraham (R-MI)	$67,620
3	John D Ashcroft (R-MO)	$438,650	John D Ashcroft (R-MO)	$188,349	John D Ashcroft (R-MO)	$442,156	John D Ashcroft (R-MO)	$186,836	Richard J Santorum (R-PA)	$455,650	Calvin M Dooley (D-CA-20)	$60,070
4	Rick A Lazio (R-NY-2)	$433,660	Rick A Lazio (R-NY-2)	$174,396	Rick A Lazio (R-NY-2)	$419,377	Conrad Burns (R-MT)	$184,628	John D Ashcroft (R-MO)	$450,719	Larry Ed Combest (R-TX-19)	$52,981
5	Conrad Burns (R-MT)	$389,538	Conrad Burns (R-MT)	$170,250	Conrad Burns (R-MT)	$365,686	Rodney Dwight Grams (R-MN)	$176,937	William V Roth Jr (R-DE)	$412,303	John D Ashcroft (R-MO)	$51,387
6	Rodney Dwight Grams (R-MN)	$377,374	William V Roth Jr (R-DE)	$157,500	Rodney Dwight Grams (R-MN)	$355,934	Richard J Santorum (R-PA)	$166,692	Rodney Dwight Grams (R-MN)	$387,665	Richard G Lugar (R-IN)	$48,500
7	William V Roth Jr (R-DE)	$364,000	Slade Gorton (R-WA)	$157,025	William V Roth Jr (R-DE)	$339,154	Rick A Lazio (R-NY-2)	$164,592	Conrad Burns (R-MT)	$385,797	Richard J Santorum (R-PA)	$44,560
8	Mike Dewine (R-OH)	$350,421	Rodney Dwight Grams (R-MN)	$154,394	Mike Dewine (R-OH)	$326,079	Anne Meagher Northup (R-KY-3)	$143,895	Slade Gorton (R-WA)	$361,870	William M Thomas (R-CA-21)	$38,569
9	Slade Gorton (R-WA)	$309,587	Mike Dewine (R-OH)	$151,321	J Dennis Hastert (R-IL-14)	$309,732	Mike Dewine (R-OH)	$141,809	Mike Dewine (R-OH)	$338,900	Rick A Lazio (R-NY-2)	$38,020
10	J Dennis Hastert (R-IL-14)	$302,331	J Dennis Hastert (R-IL-14)	$114,250	Slade Gorton (R-WA)	$288,607	William V Roth Jr (R-DE)	$140,000	J Dennis Hastert (R-IL-14)	$303,168	C Trent Lott (R-MS)	$37,500

Reprinted with the permission of the Environmental Working Group, from www.ewg.org

ple to uncover influence peddling. Like the broader Center for Responsible Politics, the Environmental Working Group tracks political action committee (PAC) money and correlates campaign contributions to the voting patterns of its recipients. The most celebrated of its databases is what it calls the Dirty Money Tracker, the environmental counterpart to Open Secrets, the website of the Center for Responsible Politics. This site monitors the campaign contributions of antienvironmental interests accepted by candidates and reveals the results by candidate, by PAC, and by state. The premise is that policy decisions are often corrupted by financial interests and that citizens should know the economic beneficiaries of policy decisions. Websites of other organizations track information such as the foods on marketplace shelves that are directly or indirectly the product of genetic modification, species moving toward endangerment or extinction, and the prevalence of persistent organic pollutants (POPs).

Another kind of political database is maintained by several environmental organizations. The most comprehensive one is that of the Defenders of Wildlife. This site monitors the introduction and disposition of legislative riders, a subject discussed at length in Chapter 2. Again, because of the relative obscurity and secrecy of these legislative stratagems, pro-environmental groups have committed special resources to exposing and publishing them.

Two characteristics of many of the databases to which people may turn for information merit emphasis, because they help drive the politics of policy development. The first is the breakdown of information geographically, so that visitors to the site can learn precisely the nature and magnitude of the threats to their own states, cities, and neighborhoods. When Massachusetts Congressman Tip O'Neill famously said that "all politics is local," he was affirming what has become conventional political wisdom. For all their generalized concern and compassion for the more serious and overwhelming environmental circumstances and conditions, most people are moved to action, if at all, by their own local—and personal—situations.

The second is that the information is usually broken down by issue— global warming, clean air, pesticides, wildlife—allowing users to pick and choose those threats of most concern them. Plugging into the environmental world through the Internet differs from joining a mainstream organization. Users need not buy into a broad agenda but can focus attention on the specific matters of greatest interest to them. Perhaps more significant, concentrating on specific issues can lead one to the sites of a number of organizations, even previously obscure ones linked to the mainstream site. Website surfing can, thus, expose the visitor to a range of information and perspectives on a subject in a way that membership in one or even a few organizations cannot. Its capacity to segregate information by geography and by subject is fundamental to the Internet's political power.

Access to the system and to information about issues is crucially related to communication. Once interest groups, or even individual citizens, know that untold numbers of people are "out there" who share their views on an issue or policy, they can contact them and join a network of people with common concerns. Such access also enables them to keep track of their opponents. *Network* is in fact the word that best describes what modern telecommunications and computers have created among their users. Conceptually, a network is a vast number of parts that are interconnected by a system of lines and routes. Individual e-mails and group e-mails, called listservs, are the principal lines and routes through which environmental organizations can expand their contact base, reaching out well beyond their membership; through which communities can connect with other localities with similar problems and share ideas and solutions; and through which individuals can find like-minded souls and develop ad hoc interest groups.

These networks have been created from both the top down and the bottom up. Users who visit the home pages of the major organizations are invited to supply their e-mail addresses or to register for listservs and to provide the same information for their friends. Less well-known or -funded interests, community groups, and institutions seeking to disseminate information or research they have generated or uncovered do the same. Individuals also can request to be added to listservs, news services, and communications networks. The power of the Internet can be used—and is being used—by individuals to join a group or start or expand one, in effect helping them to accomplish many of the same collective objectives that organizations historically have through lobbies but with relative ease and speed and at little or no cost. In short, computer technology has made possible the development of an egalitarian environmental movement and given new meaning to the process of grassroots organizing.

The Internet as Lobbying Tool

The capacity of the new technology to isolate and target issues, legislative bills, and regulations and to contact influential policymakers—in short, its lobbying capacity—has most profoundly influenced environmental politics. To be sure, informed citizens familiar with the policymaking process and knowledgeable about environmental conditions, threats, and circumstances have often contributed their views to the national environmental debate. But the voices of such people, limited in volume, have almost invariably been overwhelmed by the major players. The Internet has changed all that, in several ways.

First, simple letter writing to members of Congress has never required so little time and effort, and congressional staffs have often confirmed that such personal correspondence is being taken increasingly seriously. Numbers matter, and electronic technology generates numbers. It does so principally by on-line action networks established and maintained at websites of virtually all major environmental advocacy groups. Though they differ in details, almost all such networks users to add their names to e-mails and faxes petitioning their representatives in Congress or the president to take, or forestall, some action or policy initiative, especially in response to alerts that the advocacy groups publicize in response to emergent circumstances. These e-mails and faxes are sent collectively, adding political weight to the request. Some prepare talking points to be included in letters or even sample letters. The Rainforest Action Network has a slightly different, though interesting strategy: e-mails and faxes generated by its action network are sent directly to those it views as corporate villains: the group believes that economic pressure is more effective than political action. It even makes a special effort to reach those on the highest levels of the corporate hierarchy. The Rainforest Action Network can claim some successes, most notable of which was its motivating Home Depot, the nation's leading home improvement outlet and the largest lumber retailer in the world, to terminate sales of wood products from endangered forests by 2002.

So popular has this new cyber activism been that advocacy organizations have entered into alliances with their sister organizations and shared activist networks to supplement their own. The largest in number, if not prominence, is the Action Network, a consortium of more than 30 of the smaller environmental associations, such as Zero Population Growth and American Rivers, that enables recipients of an e-mail alert to send a fax to a public official simply by replying to the e-mail. By the middle of last year, alerts had been sent to almost half a million supporters, and the groups claimed to have registered almost 400,000 electronic activists, an impressive "community of like-minded e-mail activists," as Benjamin Smith, Environmental Defense's outreach coordinator called it.

A much more high-profile cyber network alliance is that comprising 16 of the most prominent groups and called the Save Our Environment Action Center. This coalition includes Environmental Defense, the National Audubon Society, the World Wildlife Fund, the Sierra Club, the Wilderness Society, Defenders of Wildlife, and other influential organizations. The targets of Save Our Environment are less across-the-board than significant, emergent issues worthy of campaigns. Save Our Environment takes much credit, for example, for passage of the Everglades restoration legislation of several years ago, which it attributes to its systematic network campaign supporting it.

Electronic Campaigns

"Campaigns" are important components of this new environmental strategy. The Alaska Rainforest Campaign, a project of the Alaska Conservation Foundation, has enlisted the combined efforts of groups such as the Sierra Club, Defenders of Wildlife, and the Wilderness Society in developing its own website to protect Alaska's remaining wildlands. Elements of its Internet strategy are the now familiar prewritten faxes to key governmental officials, action alerts, related links, and relevant news releases, but it also provides another, unique function—a virtual walk through the Alaskan wilderness. The capacity of the Internet to generate simulations of actual experience to dramatize verbal messages is only now starting to be used. It will no doubt become an increasingly important feature of such sites.

Another "campaign" website, "thecampaign.org," pursues the sole mission of enacting a legal requirement that foods that are the product, directly or indirectly, of genetic modification be so labeled. This site reproduces all legislation introduced on the subject, together with sponsors' statements thereon, as well as a host of other information justifying passage of the measures. It also links to websites of other organizations—Mothers for Natural Law and a variety of consumer-friendly interests—with similar objectives.

Two campaigns of a somewhat different nature were launched at the outset of the George W. Bush administration. Perhaps for the first time, an electronic campaign was initiated to defeat the nomination of a cabinet designee, Gail Norton, for secretary of the interior. Virtually all the major environmental groups testified formally at committee hearings on the nomination, but the incredible public response generated by the website "SayNoToNorton" was largely credited with playing a key role in influencing the 24 votes against her, an unusually large number for cabinet nominees. The failure of the environmental community to defeat the nomination was perhaps the inspiration for another campaign, this to "Save the Arctic Refuge from Big Oil." The position of both the new president and his secretary of interior was that the Alaskan National Wildlife Refuge could indeed be drilled for its oil reserves without undue damage to the environment, a position at the center of the environmentalist protest of her confirmation. The launching of this campaign by the Defenders of Wildlife, supported and joined by other mainstream groups, came not coincidentally only weeks after Gail Norton's confirmation and during the throes of an electric power pricing crisis in California that pro-oil interests exploited as further justification for oil exploration in Alaska. Although some of these campaigns are broad in scope and perennial in duration, recent campaigns using the resources of the Internet have followed the news and responded immediately to urgent matters in a way that the environmental community, through traditional channels, never could before.

Internet-assisted campaigns are by no means limited to domestic matters; indeed, they can surmount challenges posed by global threats that traditional organizational mechanisms cannot. For example, a campaign to support a United Nations-sponsored effort to develop a treaty to eliminate a class of chemicals called persistent organic pollutants (POPs) from the global environment is being waged by an alliance comprising the Environmental Working Group, the Environmental Health Fund, the Ecology Center, Commonweal, and the Center for Health, Environment, and Justice. The signatories to the treaty now under negotiation number more than 50 and represent more than a dozen countries. Persistent organic compounds, which include polychlorinated biphenyls (PCBs), DDT, and dioxin, have been implicated in birth defects, learning disabilities, infertility, and immune system damage and are transmitted from mother to baby during pregnancy. Given the international reach of the problem and the geographical and language diversity of the potential signatories to the treaty, such a campaign could not have been conducted practicably, if at all, without the capacities of the Internet. Indeed, it would have been much easier for the major economic interests, in this case transnational chemical companies, to control the flow of information, and, hence, the direction of any policies, in the absence of the Internet.

Tool of Environmental Revolution

The Internet, then, in the relatively short period of a decade, has revolutionized environmental politics. It has enabled individuals and environmental interests not historically represented in the halls of Congress or state capitols—interests not well funded or well known—to compete at a credible level with the imposing business conglomerates, and even national environmental groups, for political attention. It has done so by making accessible to them—at little or no cost and at great convenience—an unprecedented body of information on the processes of policymaking, proposed legislation, and the full spectrum of environmental threats, concerns, and issues. It has also allowed them access to the complex regulatory process by publicizing proposed regulations, and it has connected them to the identities and agendas of groups that share their views, as well as those that oppose them.

With the myriad news updates on the websites of the major environmental organizations, as well as on those with news as their principal purpose such as the Environmental News Network (ENN), environmental issues are no longer held hostage to the "newsworthiness" criterion that prevails in the popular media. It is very likely that more people get their environmental news from the Internet than from ABC, CBS, NBC, and CNN combined. Environmental websites speak directly to their visitors and thus open the public debate to subjects absent from the national agenda. At the same time, by providing databases that categorize information along geographical lines, these sites allow regional media to identify local concerns and threats that are of greatest interest to their audiences.

Moreover, by providing a popular audience for the vast storehouses of scientific studies that are added to daily, the World Wide Web frees citizens from the constricted parameters of advocacy science that now prevail and opens up to public debate whole new scientific perspectives on familiar problems. Scientists not in the government or the private sector who are in the field, pursuing pure science for academic institutions, foundations, and institutes, are now finding their work on the World Wide Web, unmediated by reporters, and drawn upon to support or refute policy statements or positions by cyberspace visitors.

The effects of the Internet revolution on environmental policymaking have also been felt by advocacy organizations. During the past four decades, the environmental agenda has diversified in response to a variety of forces discussed earlier, as have the advocacy organizations themselves. The dominance historically enjoyed by national organizations has been compromised, and local, grassroots, health-based, socially conscious groups have correspondingly gained political strength. A not-insignificant supplement to this organizational change has been the incorporation of groups, interests, and governmental bodies from abroad, into national initiatives. The dramatic impact of European reaction to genetically modified foods and the steady progress of a number of European and Asian nations to meet commitments made at Kyoto regarding global climate change are but two examples of the effects of foreign influence on U.S. policy. As environmental threats have increasingly become recognized as global, the association of national and international forces has accordingly become more important, substantively and politically. Communications among activists from nations thousands of miles apart would have been seriously hampered but for the Internet, which leaps oceans with as much facility as its visitors leap websites.

It is the capacity of the Internet to mobilize otherwise disparate interests and forces from the U.S. and around the world that makes it a major force in environmental policymaking. The ready availability of website action networks has resulted in an unprecedented number of communications to government officials and corporations from citizens and organizations that were previously effectively excluded from the national political debate. Lobbying, like environmental concerns themselves, has been brought home as millions of independent citizens from their homes and offices join interest group representatives patrolling the halls of Congress and state capitols in expressing their will to policymakers.

Of course, there are caveats. First, but surely not least, is the so-called digital divide. Although the Internet has gone a long way toward democratizing the political process and in expanding the number of participants dramatically, it remains an unfortunate fact that access to computers is not equal. People who own home computers are often wealthier and better educated than those who don't. That situation is, of course, changing daily, but until access to computers is uniform, access to the political policymaking arena will remain uneven, and the policies advocated in it will betray a real, though diminishing, class bias.

Second, e-mail activism may be a victim of its own success. The ease and economy of communicating to policymakers through Internet action networks has dramatically increased participation, but the conviction behind participation may be qualified by the very ease with which it is expressed. Legislative staffers have for years conceded that group actions, such as petitions, have less influence on lawmakers than do other forms of communication, especially handwritten letters, simply because they are "too easy." Internet action networks make those group communications even easier. Some lawmakers in Washington have even privately conceded that they discount huge volumes of mail from such sites and have their own technological means of segregating them. That is partly why some of the sponsors of action networks recommend—and send—faxes in lieu of or in addition to e-mails. Sending faxes is, of course, almost as easy as sending e-mails, but because a fax results in a hard copy, it is less likely to be discounted or discarded. In any case, given the realities of the modern world, it will be increasingly difficult for policymakers to ignore electronic transmissions, especially given their volume. In fact, the White House now maintains a specific office to receive and acknowledge e-mail correspondence to the President.

Third is the reality that the Internet is used as purposefully by business and corporate interests with an antienvironmental bias as it is by environmental advocacy groups and individuals. There is a fear in some circles that, as the Internet gradually assumes more political power, corporate forces will redouble their efforts to dominate the medium. Dueling websites, after all, are not much preferable to dueling scientists. For example, press releases publicized on ENN from conservative think tanks like the American Enterprise Institute, property rights advocates such as the Oregon Lands Coalition, and representatives of several extractive industries hailed the Norton appointment as a welcome return to a policy that "balances conservation and development." Nevertheless, the greatest asset of the antienvironmental community is its vast economic resources, and the Internet surely levels that playing field. Still, environmentalists must monitor the opponents' sites as well as those of allies and not feel too emboldened by their new technological tool.

Fourth, analysts of the Internet warn of the potential for what is called "group polarization." Unlike newspapers and television news outlets, websites isolate and concentrate people of similar views, and link visitors of their sites with like-minded people on others. Rather than provide perspective and balance, websites tend to promote extremism. For those interested in organizing, the web is invaluable, but it is less valuable as an educational tool to anyone not resourceful enough to visit a variety of "unlinked" sites.

A final caveat is less often recognized but important. Almost anyone can load information on the World Wide Web, many of whose sites, unlike newspaper articles or television productions, have no editors or producers to screen them for accuracy. As the volume of materials added to the World Wide Web increases, inevitably a corresponding amount of inaccurate, even irresponsible, information will be on it. This will put increasing pressure on visitors to sites to view and use materials from unnamed or unknown sources with greater caution. To a certain extent, that has always been a good practice, whatever the print, broadcast, or electronic source. The Internet, however, is called on for immediate answers, and it has an aura of authority that it has not yet earned. The Internet is a powerful tool and, as that power continues to grow, it is sure to be exploited, but it must be used with caution.

Nevertheless, the political implications of the Internet's effects are truly staggering. Committed citizens, knowledgeable about the policymaking process, armed with an unprecedented volume and variety of information about concerns most important and immediate to them, connected to others here and abroad with similar concerns, and with ready access to tools that allow them to easily and inexpensively communicate directly as individuals or as part of an ad hoc interest group with the most influential policymakers of the country are, by any measure, a potent political force. This is the very definition of empowerment.

By conferring newfound political power on the individual, the local, and the previously unrepresented or underrepresented, the Internet has fostered the metamorphosis of environmental politics from the pursuit of broad, national policies to a populist, grassroots crusade, while, paradoxically, expanding its field of vision from the domestic scene to foreign lands. Environmental politics will never be the same.

CENTRAL IDEAS

The Internet, by providing citizens ready access to policymakers, arming them with an unprecedented amount of information, and facilitating communication between and the alliance of local interests, has revolutionized environmental politics, fueling a more populist agenda and forcing the national environmental advocacy groups and major business interests to share political influence.

BIBLIOGRAPHY

SOURCES AND RESOURCES

Following is a selected list of books and articles that helped me formulate and organize many of the issues, ideas, and principles presented in this text. Some were alluded to in the book, and others provided important perspectives and background. I have also included and annotated those works that offer thoughtful counterarguments to the ones I have advanced. Individually and collectively these works may profitably serve as springboards for discussion, further study, or writing.

Chapter 1

A number of fine books trace and analyze the development of environmentalism and environmental policy. Two that I found especially enlightening and insightful, while offering sharply different critiques of the environmental movement, are Dowie, Mark, *Losing Ground: American Environmentalism at the Close of the Twentieth Century*, MIT Press, Cambridge, 1995; and Kaufman, Wallace, *No Turning Back: Dismantling the Fantasies of Environmental Thinking*, Basic Books, New York, 1994. Other useful studies are Hays, Samuel P., *Beauty, Health, and Permanence: Environmental Politics in the United States, 1955–1985*, Cambridge University Press, Cambridge, 1987; Pursell, Carroll, ed., *From Conservation to Ecology: The Development of Environmental Concern*, Thomas Y. Crowell Company, New York, 1973; and Dunlap, Riley E., and Mertig, Angela G., *American Environmentalism: The U.S. Environmental Movement, 1970–1990*, Taylor and Francis, Washington, DC, 1992.

Chapter 2

A skeletal description of the legislative process can be secured from, among other sources, the Congressional Research Service. The elements of lawmaking—legislative, regulatory, and judicial—are treated more extensively in Fiorino, Daniel, *Making Environmental Policy*, University of California Press, Berkeley, 1995, chaps. 2 and 3; Rosenbaum, Walter A., *Environmental Politics and Policy*, 3rd ed., CQ Press, Washington, DC, 1995, chap. 3; Switzer,

Jacqueline Vaughn, *Environmental Politics: Domestic and Global Dimensions*, St. Martin's Press, New York, 1994, chap. 3; and Buck, Susan, *Understanding Environmental Administration and Law*, Island Press, Washington, DC, 1991, chap. 3.

The website of the League of Conservation Voters (www.lcv.org) is the most detailed and up-to-date source of congressional proceedings, and its annual Scorecard is the best summary of the year's significant legislative activity, as well as the voting records of each senator and representative on major issues.

Chapter 3

A study of the policy implications of regulation might well start with two classics in the field that represent opposing points of view: Wilson, James Q., *The Politics of Regulation*, Basic Books, New York, 1980, esp. chaps. 8 and 10; and Tolchin, Susan J. and Martin, *Dismantling America: The Rush to Deregulate*, Oxford University Press, New York, 1983, esp. chaps. 1, 4, and 6. Other perceptive and challenging works include: Yandle, Bruce, *The Political Limits of Environmental Regulation: Tracking the Unicorn*, Quorum Books, New York, 1989; and Hoberg, George, *Pluralism by Design: Environmental Policy and the American Regulatory State*, Praeger, New York, 1992.

Two critiques of the EPA that are indispensable to an understanding of its functioning and the political pressures under which it has operated are Landy, Marc K., Roberts, Marc J. and Thomas, Stephen R., *The Environmental Protection Agency: Asking the Wrong Questions*, Oxford University Press, New York, 1990; and *Setting Priorities, Getting Results: A New Direction for the Environmental Protection Agency*, National Academy of Public Administration Report to Congress, April 1995. Shorter pieces that provide valuable perspectives on the EPA's regulatory responsibilities and history are Portnoy, Paul, EPA and the Evolution of Federal Regulation, in *Public Policies for Environmental Protection*, Portnoy, Paul, ed., *Resources for the Future*, Washington, DC, 1990; and Rosenbaum, Walter A., The EPA at Risk: Conflicts over Institutional Reform in *Environmental Policy in the 1990's*, 3rd ed., Vig, Norman J. and Kraft, Michael E., eds., CQ Press, Washington, DC, 1997.

Clinton, President Bill and Gore, Vice President Al, *Reinventing Environmental Regulation*, a National Performance Review report issued on March 16, 1995, not only sets forth the 10 principles for reinventing environmental regulation noted in the text, but also reviews the past 25 years of regulation, provides the administration's vision for the next 25 years, and enumerates 25 high-priority actions. It is the most succinct and authoritative statement of the administration's anticipated direction for the EPA and the rationale therefor.

Some of the excesses of regulation that the Clinton administration presumably sought to address are given delightfully mock treatment in Reich, Robert B., The Origins of Red Tape, *Harvard Business Review*, May/June, 1987, reprinted in *The Resurgent Liberal*, Vintage Books, New York, 1991, pp. 34–47, esp. 44–47.

As suggested throughout Chapters 4 and 5, the roles of the media and the scientific community in policymaking are intertwined. The public gets its environmental "education" principally from the media, who go to scientists for their information. The scientists that the media seek out are those who are accessible, intelligible to laypersons, and committed to the public, but relations between the two are inevitably strained. I have broken out those sources that focus principally on one or the other, but the policy implications of the synergies between the two can be fully appreciated only by reading them in tandem.

Chapter 4

A first-rate discussion of the influence of the news media on our culture and politics is contained in Schudson, Michael, *The Power of News*, Harvard University Press, Cambridge, MA, 1995, esp. chaps. 8–10. A brilliant, justifiably classic analysis of media coverage of science and technology issues is Nelkin, Dorothy, *Selling Science*, rev. ed., W.H. Freeman and Company, New York, 1995.

A unique collection of pieces on the issues, problems, and challenges of environmental journalism, principally but not exclusively written by journalists, editors, and producers, is LaMay, Craig L. and Dennis, Everette E., *Media and the Environment*, Island Press, Washington, DC, 1991. Other perspectives on environmental journalism from insiders are Prato, Lou, *Covering the Environmental Beat: An Overview for Radio and TV Journalists*, The Media Institute, Washington, DC, 1991, and Frome, Michael, *Green Ink: An Introduction to Environmental Journalism*, University of Utah, Salt Lake City, 1998. Finally, an excellent book on the history of communicating scientific issues to the public is Gregory, Jane and Miller, Steve, *Science in Public: Communication, Culture, and Credibility*, Plenum Press, New York, 1998, esp. chaps. 5–7.

To my knowledge, the best short piece on the interrelationships between science and the media with respect to risk communication is Sandman, Peter M., *Explaining Environmental Risk*, U.S. EPA, Washington, DC, 1986.

Insightful discussions of the practices and procedures involved in "producing" environmental stories are Hannigan, John A., *Environmental Sociology*, Routledge, London, 1995, esp. chaps. 2 and 3, and Hansen,

Anders, ed., *The Mass Media and Environmental Issues*, Leicester University Press, Leicester, 1993, esp. chaps. 2–4.

A representative sample of periodical literature on media coverage of the environment: Edwards, David, Can We Learn the Truth about the Environment from the Media? *The Ecologist*, January/February 1998, pp. 18–22; Gerbner, Science on Television: How It Affects Public Conceptions, *Issues in Science and Technology*, Spring 1987, pp. 109–115; Stocking, Holly and Leonard, Jennifer Pease, The Greening of the Press, *Columbia Journalism Review*, November/December 1990, pp. 37–44; Goldin, Greg and Motavilli, Jim, Is TV Going Green? *E Magazine*, January/February 1995, pp. 36–41; and Lee, Martin A. and Solomon, Norman, ... And That's The Way It Is, *E Magazine*, January/February 1991, pp. 38–43, 65–67.

A representative sample of the crusade against the media's allegedly sensationalistic and irresponsible coverage of the environment: Bailey, Ronald, *Eco-Scam*, St. Martin's Press, New York, 1993, chap. 10; Ray, Dixie Lee, *Environmental Overkill: Whatever Happened to Common Sense?* Regnery Gateway, Washington, DC, 1993, chap. 14; Wildavsky, Aaron B., *But Is It True?* Harvard University, Cambridge, MA, 1995, chap. 12; Limbaugh, Rush, *See, I Told You So*, Pocket Books, New York, 1993, chaps. 13 and 14; Kaufman, Wallace, *No Turning Back*, Basic Books, Harper Collins, New York, 1994, chap. 6; and Graber, Doris, *Mass Media and American Politics*, CQ Press, Washington, DC, 1980, pp. 171–174. The most authoritative counterattack is that of Ehrlich, Paul R., *Betrayal of Science and Reason: How Anti-Environmental Rhetoric Threatens Our Future*, Island Press, Washington, DC, 1996, chap. 11.

Chapter 5

An explanation of the scientific process and useful background to many of the issues raised in this book is Carey, Stephen S., *A Beginner's Guide to Scientific Method*, Wadsworth Publishing Company, Belmont, CA, 1994.

To gain a historical perspective on the relationship between politics and science, I recommend the old standby: Greenberg, Daniel S., *The Politics of Pure Science*, New American Library, New York, 1967. Another historical perspective on science that illuminates contemporary issues is LaFollette, Marcel C., *Making Science Our Own: Public Images of Science 1910–1955*, University of Chicago Press, Chicago, 1990.

The alternately symbiotic and adversarial roles of scientists and the journalists in contemporary policymaking are accorded extensive treatment in Hartz, Jim and Chappell, Rick, *Worlds Apart: How the Distance Between Science and Journalism Threatens America's Future*, First Amendment Center, Nashville, 1997.

A formal attack on governmental regulation, albeit in the form of exploratory hearings, was conducted by Congress in 1995, and the testimony was compiled in two volumes under the rubric of *Scientific Integrity and Public Trust: The Science Behind Federal Policies and Mandates: Hearings Before the Subcommittee on Energy and Environment of the Committee on Science*, U.S. House of Representatives, 104th Congress, 1st session, held on September 20, 1995 (No. 31) and December 13, 1995 (No. 39), on stratospheric ozone and dioxin, respectively.

So anathema to his understanding and principles was the thinly disguised purpose of the hearings and its tenor, and many of the views welcomed by committee members, that Representative George E. Brown, Jr., the committee's ranking minority member, issued a dissenting report: *Environmental Science Under Siege: Fringe Science and the 104th Congress*, A Report to the Democratic Caucus of the Committee on Science, U.S. House of Representatives, October 23, 1996, reprinted in *Environment*, March 1997 (dissenting letters and Rep. Brown's responses thereto are in *Environment*, May 1997). Rep. Brown had long been interested in bringing sound science into environmental policymaking; see, for example, Brown, George, Science's Real Role in Policy-Making, *Chemical and Engineering News*, May 31, 1993, pp. 9–11.

Despite Rep. Brown's objections, the hearings ultimately resulted in the submittal of a new proposal for how science should be incorporated into environmental policymaking: *Unlocking Our Future: Toward A New Science Policy*, A Report to Congress by the House Committee on Science, September 1998.

The latest effort of the federal government to improve the scientific basis for environmental decision making was the creation of the National Council for Science and the Environment, which issued its Strategic Plan in June 2000. The plan provides a protocol for bringing stakeholders and decision makers together and providing them with access to "comprehensive, non-partisan science-based information about the environment."

A recent study of how science has been developed and applied over the years at the EPA is valuable from both policy and programmatic perspectives: Powell, Mark. R., *Science at EPA: Information in the Regulatory Process*, Resources for the Future, 1999.

For most of the 1990s, discussions over how to most effectively and responsibly apply science in environmental policymaking were taking place across the political spectrum. An indispensable discussion of the problems and some suggested solutions to them is: Ruckelshaus, William D., Risk, Science and Democracy, *Issues in Science and Technology*, Spring 1985, pp. 19–38. A balanced, nonpartisan view is offered in Wilson, James D. and Anderson, J.W., What the Science Says: How We Use It and Abuse It to Make

Health and Environmental Policy, *Resources*, Resources for the Future, Summer, 1997, pp. 5–8. The scientific community called repeatedly, and not surprisingly, for more input from scientists, e.g., Madia, William J., A Call for More Science in EPA Regulations, *Science*, October 2, 1998, p. 45.

Critiques of how science has been abused by industry and government alike abound. For a detailed attack on industry, see Fagin, Dan and Lavelle, Marianne, *How the Chemical Industry Manipulates Science, Bends the Law, and Endangers Your Health*, Common Courage Press, New York, 1998. Attacks on government's alleged misuse of science have been supported principally by conservative and libertarian foundations. Two book-length studies are among the most prominent: Bolch, Ben and Lyons, Harold, *Apocalypse Not*, Cato, Washington, DC, 1993, and Fumento, Michael, *Science Under Siege: Balancing Technology and the Environment*, William Morrow & Company, New York, 1993.

A significant body of literature has been building up over the issue of the scientific literacy of the public, which complicates the incorporation of science into policymaking. A perceptive study of some historical roots of the phenomenon is Miller, Eliana Beth, *Public Alienation from Science and Science Illiteracy: Post-Sputnik Education and the Media as Causative Agents*, unpublished senior thesis, Swarthmore College, 1993. For an extended treatment see Zimmerman, Michael, *Science, Nonscience, and Nonsense: Approaching Environmental Literacy*, Johns Hopkins University Press, Baltimore, 1995, and Howell, Dorothy, *Scientific Literacy and Environmental Policy: The Missing Prerequisite for Sound Decision Making*, Quorum Books, New York, 1992. An interesting editorial on the subject is Maienschein, Jane, Scientific Literacy, *Science*, August 14, 1998.

The problems that courts have had with evaluating the legitimacy of regulation based on science are comprehensive and thoughtfully treated in two long articles: Abraham, Kenneth S. and Merrill, Richard A., Scientific Uncertainty in the Courts, *Issues in Science and Technology*, Winter 1986, pp. 93–107; and Burack, Thomas S., Of Reliable Science: Scientific Peer Review, Federal Regulatory Agencies, and the Courts, *Virginia Journal of Natural Resources Law*, 7:27, 1987, pp. 27–110.

The literature on the precautionary principle is only just starting to grow, but Tickner, Joel and Raffensperger, Carolyn, *The Precautionary Principle in Action: A Handbook*, Science and Environmental Health Network, 2000, remains one of the best explanations of its theory and potential applications.

Chapter 6

The literature on state and local governments as political interests is scant. The standard volume on the subject is Harrigan, John J., *Politics and Policy in States and Communities*, 4th ed., HarperCollins Publishers, New York, 1991, esp. chaps. 1 and 16.

Other, briefer treatments are in Switzer, Jacqueline Vaughn, *Environmental Politics: Domestic and Global Dimensions*, St. Martin's Press, New York, 1994, pp. 67–69; Daly, Herman, Goodland, Robert and Cumberland, John H., *An Introduction to Ecological Economics*, Costanza, Robert, ed., St. Lucie Press, New York, 1997, chap. 4; Lester, James P. ed., *Environmental Politics and Policy: Theory and Evidence*, Duke University, Chapel Hill, 1995; and Davis, Sandra K., Fighting over Public Lands: Interest Groups, States, and the Federal Government, in Davis, Charles, ed., *Western Public Lands and Environmental Politics*, Westview Press, Boulder, CO, 1997.

One can track the positions on environmental issues of states and localities best by routinely consulting their respective publications and websites: *State Legislatures* by the National Council of State Legislatures; *State Government News* and *Eros*, by the Council of State Governments and its publication, *Eros*; The American Legislative Exchange Council, and *Governing: The Magazine of States and Localities*.

Chapter 7

There are any number of works that trace the history of environmentalism and its advocacy groups. Among the best are Gottlieb, Robert, *Forcing the Spring: The Transformation of the American Environmental Movement*, Island Press, Washington, DC, 1993; and Shabecoff, Philip, *A Fierce Green Fire: The American Environmental Movement*, Hill & Wang, New York, 1993.

A study of the ideological roots of environmental justice can be found in Smith, James Noel, ed., *Environmental Quality and Social Justice in Urban America*, Conservation Foundation, Washington, DC, 1974. The contemporary environmental justice movement has received extensive treatment. Among the more prominent studies are Bullard, Robert D., *Unequal Protection: Environmental Justice and Communities of Color*, Sierra Club Books, San Francisco, 1994; Schwab, Jim, *Deeper Shades of Green: The Rise of Blue-Collar and Minority Environmentalism in America*, Sierra Club, San Francisco, 1994; Foreman, Christopher H., Jr., *The Promise and Peril of Environmental Justice*, Brookings Institution Press, Washington, DC, 1998; and Bryant, Bunyan, ed., *Environmental Justice: Issues, Policies, and Solutions*, Island Press, Washington, DC, 1995.

Two governmental issue briefs are Durett, Dan, *Environmental Justice: Breaking New Ground*, 2nd ed., National Institute for the Environment, 1994; and Cooper, Mary H., Environmental Justice, *CQ Researcher*, Washington, June 19, 1998. A brilliant academic treatment is Tesh, Sylvia N.

and Williams, Bruce A., Identity Politics, Disinterested Politics, and Environmental Justice, *Polity*, Spring 1996.

Environmental organizations have, of course, weighed in. See, for example, Russel, Dick, Environmental Racism, *Amicus*, Spring 1989, pp. 22–32; and Brough, Holly, Minorities Redefine "Environmentalism," *World Watch*, September/October 1990, pp. 5–8.

The business community's response can be gleaned from: Warner, David and Worsham, James, EPA's New Reach, *Nation's Business*, October 1998, pp. 12–19; Popeo, Daniel, Environmental Injustice, *New York Times*, August 24, 1998; Litvan, Laura M., Fighting Racism: EPA's New Role? *Investors Business Daily*, March 25, 1998; and Geyelin, Milo, Pollution Suits Raise Charges of Racism, *Wall Street Journal*, October 29, 1997.

Ecoterrorism has understandably been accorded less attention, but see: Sullivan, Robert, The Face of Eco-Terrorism, *New York Times*, December 20, 1998, pp. 46–49; Markels, Alex, Backfire, *Mother Jones*, March/April 1999, pp. 60–64, 78–79; and Russell, Dick, The Monkeywrenchers, *Amicus*, Fall 1987, pp. 28–41.

Chapter 8

For a good study on the new thinking going on in the business community with respect to environment, see Hawken, Paul, *The Ecology of Commerce: A Declaration of Sustainability*, Harper Collins, New York, 1993. See also Kirkpatrick, David, The Environment: Business Joins the New Crusade, *Fortune*, February 12, 1990, pp. 44–52.

The bible of the Wise Use movement is Gottlieb, Alan M., ed., *The Wise Use Agenda*, Free Enterprise Press, Bellevue, WA, 1989. See also Arnold, Ron, *Ecology Wars: Environmentalism As If People Mattered*, Free Enterprise Press, Bellevue, WA, 1993 and Arnold, Ron and Gottlieb, Alan, *Trashing the Economy: How Runaway Environmentalism Is Wrecking America*, Free Enterprise Press, Bellevue, WA, 1994.

Three book-length studies of the antienvironment movement are: Switzer, Jacqueline, Vaughn, *Green Backlash: The History and Politics of Environmental Opposition in the U.S.*, Lynne Rienner Publishers, Boulder, CO, 1997; Rowell, Andrew, *Green Backlash: Global Subversion of the Environmental Movement*, Routledge, London, 1996; and Helvarg, David, *The War Against the Greens*, Sierra Club, San Francisco, 1994. A small, canny, but eye-opening directory of antienvironmental organizations, their sources of support, and their missions is Deal, Carl, *The Greenpeace Guide to Anti-environmental Organizations*, Odonian Press, Berkeley, CA, 1993. Short but useful pieces are Helvarg, David, Grassroots for Sale, *Amicus*, Fall

1994, pp. 24-29 and Whose Agenda for America, *Audubon*, September/October 1992, pp. 78-91.

The increasing political power of foundations at the federal and local levels is treated in Ruben, Barbara, Getting the Wrong Ideas, *Environmental Action*, Spring 1995; and Enos, Gary, The Politics of Persuasion: Conservative Think Tanks Wield Growing Influence, *City and State*, April 20, 1992. Foundation efforts to influence Congress, notwithstanding their tax-exempt status, are exemplified by Adler, Jonathan H. et al., ed., *Environmental Briefing Book*, Competitive Enterprise Institute, 1996; and Crain, Edward and Boaz, David, *The Cato Handbook for Congress*, Cato Institute Books, Washington, DC, 1995.

Brief but informative articles on corporate codes of conduct are Nash, Jennifer and Ehrenfeld, John, Code Green, *Environment*, January/February 1996, pp. 16-20, 36-45; Bavaria, Joan, An Environmental Code for Corporations, *Issues in Science and Technology*, Winter 1989-90, pp. 28-30; and Rayport, Jeffrey and Lodge, George C., Responsible Care, *Harvard Business School*, March 18, 1991.

On the trend to market "green" products see Nixon, Will, The Color of Money: Cashing in on Green Business, *Amicus*, Summer 1996, pp. 16-18; and Reisner, Marc, Green Expectations: Making Money the Environmental Way, *Amicus*, Summer 1998; as well as Weber, Peter, Green Seals of Approval Heading to Market, *World Watch*, July/August 1994, pp. 19-23.

Greenwashing is discussed extensively in Athanasiou, Tom, *Divided Planet: The Ecology of Rich and Poor*, University of Georgia Press, Athens, 1998; and in the format of an exposé in Bruno, Kenny, *The Greenpeace Book of Greenwash*, Greenpeace International, Washington, DC, undated.

Chapter 9

Three outstanding recent books on the environmental policy consequences of globalization are: Karliner, Joshua, *The Corporate Planet: Ecology and Politics in the Age of Globalization*, Sierra Club Books, San Francisco, 1997; Conca, Ken and Dabelko, Geoffrey, D., eds., *Green Planet Blues*, Westview Press, Boulder, CO, 1998; and Hilary French, *Vanishing Borders: Protecting the Planet in the Age of Globalization*, W.W. Norton, New York, 2001.

A sample of shorter pieces on the subject include: Vogel, David, International Trade and Environmental Regulation, in Vig, Norman J. and Kraft, Michael E., eds., *Environmental Policy in the 1990's*, 3rd ed., CQ Press, Washington, DC, 1997; Faux, Jeff, Slouching Toward Seattle: The WTO Behemoth: Will Trade Overwhelm Democracy—or Stimulate a New Global Politics? *The American Prospect*, December 6, 1999; Adams, Bob, World

Trade: Not Easy Being Green, *National Voter*, March/April 2001, pp. 3-7; and Hileman, Bette, The WTO and the Environment, *Chemical and Engineering News*, November 2, 1998, pp. 17-18. More openly critical of the WTO and its impact on the environment are: Shrybman, Steven, The World Trade Organization: The New World Constitution Laid Bare, *The Ecologist*, July 1999, pp. 270-275; French, Hilary, Challenging the WTO, *World Watch*, November/December 1999, pp. 23-27; Frazer, Phillip, Fighting to Stop the WTO from Trashing Environmental Laws, *News on Earth*, January 2000; as well as several essays in Mander, Jerry and Goldsmith, Edward, eds., *The Case Against the Global Economy*, Sierra Club Books, San Francisco, 1996.

Several of the websites of the major environmental advocacy groups as well as that of Public Citizen, www.citizen.org., maintain a steady watch on global developments.

Chapter 10

A challenging academic discussion of the impact of the technological revolution on the individual's participation in environmental politics is Castells, Manuel, *The Power of Identity: The Information Age—Economy, Society, and Culture*, Blackwells, London, 2000, chap. 3.

Brief discussions of the uses of the Internet for environmental goals and their perils are Newton, James W. and Rohwedder, W.J., Environmental Computer Networking: Dialing Locally, Acting Globally, *E Magazine*, March/April 1990, pp. 45-47; and Thayer, Ann M., Information Overload? *Chemical and Engineering News*, July 13, 1998, pp. 23-28.

An extended discussion of the practical applications of the Internet in environmental advocacy is Price, Tom, *Cyber Activism: Advocacy Groups and the Internet*, Foundation for Public Affairs, Washington, DC, 2000.

Index

A

Action Network, 146
Andalex Resources, 26
Administrative Procedure Act, 35
"adopt a ..." program, 97
Advancement of Sound Science Coalition, 69
advertising, environmental, 124-125
advocacy roles
  ecoterrorism, 107-109
  environmental justice, 97-105
  governments, 88-90
  history of conservationists, 91-93
  holistic and global, 107
  organizations, 18
  use of websites, 137-151
agendas
  control of legislation, 19
  "domestication," 15, 109
  "dream" agenda, 114
  Internet limiting broad agendas, 144
  media, 64
  promoted by websites, 137-151
  Reagan administration, 113-115, 121
  revised environmental, 92-94, 109
Agent Orange, 98
air pollution, 15, 71
Air Pollution Control Act of 1955, 15
Alar brouhaha, 57, 76
Alaska, 3, 27, 146-147
Alaska Conservation Foundation, 146
Alaska Rainforest Campaign, 146
Albright, Madeleine, 135
Alcoa, 119
Alford, Harry, 103
alleged risks. *See* Precautionary principle
Alliance for Safe and Responsible Lead Abatement, 3

Alliance for Sustainable Jobs and the Environment, 4, 94
alliances, for survival, 9
amenity environmental concerns, 97
American Academy of Pediatrics, 107
American Association for the Advancement of Science, 74, 123
American Electric Power Service Corporation, 112
American Enterprise Institute, 74, 150
American Farm Bureau Federation, 112, 118
American Petroleum Institute, 112
American Public Health Organization, 79
American Rivers, 146
Animal Liberation Front, 108
animal rights, 8, 107-109
antienvironmentalism, 23-25, 60, 63-64, 117-120, 160
Antiquities Act of 1906, 26-29
Apocalypse Not, 75
appropriations bills, 22
appropriations committee, 16, 24
"Are We Scaring Ourselves to Death?", 63
Arnold, Ron, 63
Arctic National Wildlife Refuge, 125
Association of American Railroads, 112
aura of responsibility, 117, 120

B

Babbitt, Bruce, 28, 69
Baden, John, 75
Bailey, Doug, 137
Bailey, Ronald, 63, 75, 120
Balling, Robert, 74
Bavaria, Joan, 116
Betrayal of Science and Reason, 75
Bhopal, India, 3, 38, 54, 115

bibliography, 153-162
Bierce, Ambrose, 1
big business, 11
von Bismarck, Otto, 13
Bituminous Coal Operators Association, 4
Bolch, Ben, 75
bovine growth hormone, 76
breast implants, silicon, 76
British Petroleum, 123
Brower, David, 93-94, 106
Brown, George E., Jr., 70-71, 157
Browner, Carol, 49
brownfields, 36, 103
bubble-policy, 96
burden (of proof), 14, 82
burden (regulatory), 37, 96, 113
Bureau of Land Management, 20, 88, 114
Burford, Anne Gorsuch, 45, 114, 115
Bush administration (1988-1992), 45, 47
Bush administration (2000-2004), 89, 125, 131
Business Charter for Sustainable Development, 116
business-driven environmental policy
    environmental advertising, 124-125
    "green products," 124, 161
    image vs. actions, 117, 120
    lobbying through surrogates, 117-120
    media involvement, 119-120
    posturing, 112, 124
    Reagan administration, 113-115
The Business Roundtable, 3
Byrd, Robert, 24
Byzantine legislative process, 21-29

C

CAFE standards, 25
California Desert Protection Act, 23
campaign contributions, 142-143
campaigns, electronic, 146-147
cancer cluster at Woburn, 98
"Can Man Save This Fragile Earth?", 55
carbon dioxide, 130
Carson, Rachel, 6-7, 15, 18, 54, 109
Carter administration, 43-44
Cascade-Siskiyou National Monument, 28-29
Castells, Manuel, 81
Cartegena Protocol on Biosafety to the Convention on Biological Diversity (2000), 79

Cato Institute, 74, 119, 120
cattlemen, 4
CCHW. *See* Citizens Clearinghouse for Hazardous Waste (CCHW)
celebrity involvement, 57-58
Census (2000), effect on interests, 84
Center for Health, Environment and Justice, 98, 107, 109, 147
Center for Media Education, 107
Center for Responsible Politics, 143
Center for the Defense of Free Enterprise, 118
Central Ideas summaries
    business-driven environmental politics, 127
    changes in environmental politics, 12
    environmental advocacy, 110
    environmental regulation, 51
    international environmentalism, 136
    legislation, 29
    media, 64
    online activism, 151
    science, 80
    state and local governments, 90
CERES Coalition, 116
CFCs. *See* Chlorofluorocarbons (CFCs)
Chamber of Commerce, 112, 116
Chemical and Engineering News, 69
Chemical Industry Council of New Jersey, 2
Chemical Manufacturers Association, 112, 115-116
Chernobyl, 16, 33, 54
Chevron, 118, 123
children's cancer cluster, 98
Chlorofluorocarbons (CFCs), 39
Citizens Clearinghouse for Hazardous Waste (CCHW), 98, 107
Citizens for the Environment, 119
civilian activists, 97-109, 137-151
civilian nuclear waste, 4
civil rights, 8, 98
Civil Rights Act, 100
Clean Air Act, 2, 32, 33, 87, 133
Clean Water Act, 87, 89, 142
climate change treaty, 25, 62, 131, 149
Clinton administration
    Byzantine practices, 21
    Congressional Review of Agency Rulemaking, 40
    Council on Competitiveness, 45
    DiCaprio interview, 58

election impact on EPA, 47
environmental legacy, 26-29, 89
environmental stewardship expected from, 47, 105
logging issues, 22-23
national monuments, 28
negotiated rule-making, 41
regulation excesses, 155
Reinventing Government principles, 49
rule by "decree," 25-29
World Trade Organization, 135
"closet greenies," 62
"clubby" relationships, 96
Coalition for Environmentally Responsible Economics, 116
Coalition of Northeast Governors (CONEG), 86
coal mining rider, 24
coastal areas, 15
Coca-Cola, 119
codes of conduct, 115-117, 161
CO_2 emissions, 131
coffee-table books, 93-94
collaboration, 87, 123
committee consideration process, 20
Commoners, Barry, 75
Common Sense Initiative (CSI), 49
Commonweal, 147
community activists, 97-105, 107-109, 137-151
The Competitive Enterprise Institute, 111, 120
complexity of issues, 59-60, 66, 68
conditional veto, 17
CONEG. *See* Coalition of Northeast Governors (CONEG)
conference reports, 16-17
conflict, newsworthiness criterion, 56-57
Congress
    business-driven environmental policy, 120-126
    legislation process, 16, 81
    media's influence, 10, 18, 21, 62
    public view as government, 82
    relationship with science, 68-72
    104th, 22, 40, 46, 50, 69, 83
    105th, 24
    106th, 50
    websites of members, 138
Congressional Accountability for Regulatory Information Act of 1999/2000, 50

Congressional Research Service, 153
Congressional Review of Agency Rulemaking, 40
conservation
    changes, 92-93
    land use control, 87-90
    origins, 5-6, 91-92
Conservation Foundation, 91
conservatives, 1
Consumer Alert, 119
Contract with America, 22, 47-48, 69, 83, 122
Co-op America, 9
Coors, Joseph, 119
"corporate average fuel economy" standards, 25
corporate codes of conduct, 115-117, 161
corporate donor support of groups, 95-96
corporate environmentalism, 121-123
cost-benefit analysis
    as adoption precondition, 114
    Contract with America, 69
    emission limit economic burden, 96
    General Accounting Office involvement, 51
    major legislation, 50
    public policy integration, 44
    Reagan administration agenda, 115
    sound science commitment, 70
Costle, Douglas, 43
Council of State Governments, 85
Council on Competitiveness, 45-46
courts, relationship with science, 72-73, 158
Craig, Larry, 27
crises
    effect on business agenda, 115
    effect on legislation, 16, 18, 20
    media coverage, 53-64
critique of culture, 105-107
cruise ships, 8
CSI. *See* Common Sense Initiative (CSI)
cultural critique, 105-107
cyberactivism, 137-151

D

Daley, Richard M., 2
D'Amato, Alfonse, 2
Danson, Ted, 57
DDT, 6, 18, 147
Death of Common Sense, 48

decentralization of power, 21, 115
decision basis, 72, 85
"Declarations of Goals and Policy," 16
Defenders of Wildlife, 25, 91, 144, 146
Defense of the Environment Act, 24
deliberative process limitations, 20
Democrats, 1
Department of Agriculture, 24
Department of Commerce, 24
Department of Defense, 24
Department of State, 24
Department of the Interior, 19, 24
Department of Transportation, 24
developers, 4, 7
development of environmentalism, 153
devolution, 45, 83
diapers, 77
DiCaprio, Leonardo, 58
Dicks, Norman, 24
digital divide, 149
Dingell, John, 2, 31
dioxin, 77, 98, 147
Dirty Money Tracker, 143
disasters. *See* Crises
Dole, Bob, 47
dolphins, 133
"domestication" of agenda, 15, 109
Doolittle, John, 70
Downs, Anthony, 103
DuPont, 39

E

Earth Day, 125, 130, 135
Earth First!, 108
Earth Island Institute, 94, 107
Earth Justice Legal Defense Fund, 92
Earth Liberation Front, 108
Earth 911 Promotions Group, 58
eco-labeling of packaging, 132
Ecologist, 107
Ecology Center, 147
economic boom, post-World War II, 14-15,
    93, 113
ECOS. *See* Environmental Council of the
    States (ECOS)
*Ecoscam: The False Prophets of Ecological
    Apocalypse*, 120
ecoterrorism, 107-109, 160
education, from media, 154
Ehlers, Vernon, 71

Ehrlich, Paul, 75
electromagnetic fields, 15
electronic campaigns, 146-147
Elsaesser, Hugh, 74
e-mail, 144, 145, 149
emission standards, 87
"Endangered Earth," 55
endangered species, 15, 24, 118, 133
The Endless Frontier, 71
endocrine disrupters, 76
ENN. *See* Environmental News Network
    (ENN)
Envirofacts Warehouse website, 139
"The Environment: Business Joins the New
    Crusade," 121-123
Environmental Council of the States
    (ECOS), 86
Environmental Defense, 140, 146
Environmental Defense Fund, 58, 92, 94,
    95, 97, 123, 140
*Environmental Diplomacy: The
    Environment and U.S. Foreign
    Policy*, 135-136
Environmental Equity Workgroup, 100
*Environmental Gore: A Constructive
    Response to Earth in the
    Balance*, 75, 120
environmental groups. *See also* specific
    organization
    change in focus, 92-93
    exclusion during Reagan administration,
        114
    funding, 95-96
    history and development, 91-92, 153, 159
    Internet effect, 139-151
    politicization of, 94-97
    relocation trend, 96-97
Environmental Health Fund, 147
Environmental Industry Coalition of the
    United States, 3
environmentalism, 91-92, 159-160
environmental justice
    application of, 102-105
    Civil Rights Act roots, 100
    Clinton administration impact, 99
    history, 97-102
Environmental Justice Transition Group,
    100
Environmental Media Association, 58
Environmental Monitoring for Public Access
    and Community Tracking, 139

Environmental News Network (ENN), 148
Environmental Overkill, 75
Environmental Protection Agency (EPA)
   antienvironmental riders, 24-25
   Civil Rights Act, 100
   conflict of mandates, 103
   decision making by states, 85
   early history, 42-44
   environmental justice, 99
   functioning of, 154
   minimum national standards, 82
   open door for business, 114
   pressures operated under, 154
   proposed budget cut, 24
   Reagan administration impact, 44-48
   redefining focus, 69, 121
   scapegoat status, 44
   scientific capacity, 68, 72
   undermining by WTO, 129-136
   websites, 139
Environmental Science Under Siege: Fringe
    Science and the 104th
    Congress, 70-71
Environmental Working Group, 140-143,
   147
EPA. *See* Environmental Protection Agency
   (EPA)
ethanol production, 88
ethics, 8, 18-19
Exaggerated risk stories, 63-64. *See also*
    Exploitation
Executive Orders
   12291, 44, 114
   12898, 49, 99
Exploitation, 75, 80, 108. *See also*
    Exaggerated risk stories
Exxon, 118

F

faxes as lobbying tool, 145, 149
Federalism, 45, 50, 83, 86
Federalism Act of 1999, 50
"federal mandate, federal pay," 47, 83, 121
federal policies, 37, 82-86, 105
Federal Water Pollution Control Act (1948),
   15
First Amendment Center, 73
The First International Conference on
    Protection of the North Sea
    (1984), 78

First National People of Color Leadership
    Summit, 49, 100
fish stock, 123
food
   additives, 15
   genetically modified, 12, 79, 131-132
Food and Water, 107
The Food Quality Protection Act of 1996, 18
Foreman, Dave, 108
forests, 3, 8, 23, 57, 107-109, 145
Forest Stewardship Council, 123
Fortune, 121
Foundation for Research on Economics
    and the Environment, 119
Foundation on Economic Trends, 107
foundation support of groups, 95
Fox News website, 76
Frankenstein, portrayal of scientists, 77
free enterprise promotion, 74
free trade, 129-136
Friends of the Earth, 94
fuel economy standards, 24-25, 123
Fumento, Michael, 63, 66, 75
funding
   appropriation for environmental law, 14
   compliance cost, 68
   institutional interests, 20

G

Gaebler, Ted, 48
gas deposits, methane, 4
gasoline standards, 133
GATT, 62
General Accounting Office (GAO), 51
General Motors, 116, 123
genetically modified foods, 12, 79, 131-132,
   143, 146
Gibbs, Lois, 91, 98
Gifford, Bill, 91
Gingrich, Newt, 22, 40, 69, 71
Glickman, Dan, 28-29
global advocacy, 107
Global Environmental Management
    Initiative (GEMI), 116
globalization, 129-136, 161-162
global warming, 58, 71, 123, 130
golf courses, 8
Gore, Al
   Cascade-Siskiyou National Monument,
    28

Earth in the Balance, 106
election impact on EPA, 47
free trade policies, 135
global climate change appeal, 131
Handford Reach National Monument, 29
Reinventing Environmental Regulation,
41-42, 154
Reinventing Government principles, 49
stewardship expected from, 47, 105-106
Gorsuch, Anne, 45, 114, 115
Gottlieb, Alan, 63
government agencies as lobbyists, 19, 81-90
Grand Staircase-Escalante National
Monument, 26
"great facilitator," 12
"green" and profitable, 123, 161
Greene, Enid, 27
greenfields, 36
greenhouse gases, 123, 130
Greenpeace, 125
"green" products, 124, 161
"greenwashing," 124, 161
gridlock, 21
grocery bags, 77
group polarization, 150
guardianship of environmental concerns, 97

H

Hammonds, Tim, 65
Hanford Reach National Monument, 29
Hansen, James, 27
Harrelson, Woody, 57
Hatch, Orrin, 27
health threats, public, 6-7, 15, 61, 97-105
Heritage Foundation, 44, 74, 119
Hetch Hetchy, 6, 92
holistic advocacy, 107
Home Depot, 123, 146
House of Representatives
Contract with America, 47
legislation process, 16-17, 19
omnibus bill, 23
House Science Committee, 71
Howard, Philip, 48
HR9 flow chart, 122
Huber, Peter, 75
Humane Society, 107

I

IBM, 119
illegal practices, 18-19
image of responsibility, 117, 120
incidents. *See* Crises
indoor air pollution, 7, 15, 61
industry constituencies, 83-84
information technology. *See* Internet
Institute, West Virginia, 116
Interagency Working Group on
Environmental Justice, 99
interest groups. *See* Lobbyists
international concerns, 11-12, 129-136, 148
International Monetary Fund, 135
Internet. *See also* Websites
accuracy of information, 150
caveats of, 149-150
Congress members websites, 138
digital divide, 149
electronic campaigns, 146-147
geographical information, 144
issue segregation, 144
lobbying tool revolution, 12, 138,
147-151, 162
issues, complexity of, 59-60, 66, 68
Izaak Walton League of America, 91

J

Job Creation and Wage Enhancement Act,
47
"junk science," 73, 75-76

K

Kosovo emergency funding bill, 22
Kyoto treaty, 4, 12, 25, 62, 131, 149

L

labeling of packaging, 132
land trusts, 4
land use, 14, 33, 87-90, 114-115, 121
Latham, Earl, 31
lawmaking process. *See also* Legislation
changes, 9-10
regulation comparison, 32, 34
relationship with science, 66-68
lead, 15
League of Conservation Voters, 25, 154

Lee, Charles, 99
Left wing, 1
legal system, 72-73
legislation. *See also* Lawmaking process
　budget approval, 40
　crises effect, 16, 18, 20
　history, 14-16
　monument designations, 25-29
　process, 16-20, 153
　regulation comparison, 32, 34
　riders, 21-25
"legislative intent," 39, 40
legislators, behavior, 37-38
letters as lobbying tool, 145, 149
liberals, 1
Limbaugh, Rush, 63, 75
Lindzen, Richard, 74
Listservs, 144
Lobbyists
　government agencies as, 19, 81-90,
　　158-159
　Internet as tool, 145-151
　legislators as, 40
　media as, 64
　public acceptance, 1
　regulations involvement, 38-39
　surrogates, 117-120
　technology, 145-151
　using media, 60
Local governments, 11, 81-90, 105, 158-159
locally undesirable land uses (LULUs), 104
Lockheed, 119
logging industry, 3, 22-23, 25, 88
Louis-Dreyfus, Julia, 58
Love Canal, 16, 44, 54, 98
lowest common denominator standards,
　133
low-income concerns, 48
LULUs (locally undesirable land uses), 104
Lyon, Harold, 75

M

Maastricht Treaty of the European
　Community (1994), 79
Making Peace with the Planet, 75
Mandate for Leadership, 119
Mandates Information Act of 1999, 50
Marine Stewardship Council, 123
market-driven environmental policy
　environmental advertising, 124-125

"green" products, 124, 161
　image vs. actions, 117, 120
　lobbying through surrogates, 117-120
　media involvement, 117
　posturing, 112, 124
　Reagan administration, 113-115
Massachusetts Toxic Use Reduction Act
　　(1990), 78
mass mailings, support of groups, 96
McDonald's, 97, 123
media
　absence in regulation matters, 38, 39
　"construction" of stories, 58-63
　influence, 10, 18, 21, 62, 155-156
　as interest group, 64
　newsworthiness, 56-57, 73, 148
　New York Times, 25, 48, 75, 94, 106, 107
　obligations, 54
　public education, 155
　public view toward, 55-56
　science, relationship with, 73-78, 156
　self-view, 53
　sources, 59-60
　ventriloquism of scientists, 76
Michaels, Patrick, 74
Milloy, Steven, 75
mineral rights, 4
minimum national standards, 82
mining concerns, 24, 26-27, 118
minority concerns, 8, 48, 99-105
Mobilization for Global Justice, 135
Mohave National Preserve, 23
Monsanto, 119
Monsanto pact, 62
Montreal Protocol on Substances That
　　Deplete the Ozone Layer, 79
monument designations, 25-29
Mothers for Natural Law, 147
Motor Vehicle Manufacturers Association,
　　112
Mountain States Legal Foundation, 89, 120
movies (environmental messages), 58
MTBE, 77
Muir, John, 5, 93
Multiple Use Strategy Conference, 118
municipality governments. *See* Local
　governments
Murkowski, Frank, 27
Muskie, Edmund, 42

N

NAFTA, 62, 126, 130
National Academy of Public Administration, 46
National Academy of Sciences, 29
National Association of County Officials, 85
National Association of Manufacturers, 112
National Audubon Society, 91, 95, 146
National Biologic Service, 69
National Black Chamber of Commerce, 103
National Council of State Legislators, 85
National Council on the Environment, 72
National Environmental Performance Partnerships (NEPPs), 85
National Environmental Policy Act, 27
National Geographic, 55
National Governors' Association, 85, 87, 107
National Institute for the Environment, 71-70
National League of Cities, 85
National Oceanic and Atmospheric Administration, 69
National Parent Teacher Association, 107
National Parks and Conservation Association, 91, 95
National Park Service, 23
National Science Foundation, 71-70
National Wildlife Federation, 91
Natural Awakenings, 9
Natural Resources Defense Council, 25, 92, 94, 95
Nature Institute, 107
negotiated rule-making (regneg), 41
Nelkin, Dorothy, 76
NEPPs. *See* National Environmental Performance Partnerships (NEPPs)
Networks (Internet), 144
New Federalism, 83
New Republic, 55
news. *See* Media
news services (online), 144
newsworthiness, 56-57, 73, 134, 148
The New Yorker, 54
New York Times, 25, 48, 75, 94, 106, 107
NIMBY ("not in my backyard"), 8, 104
Nixon administration, 42
NoMoreScares.com, 75
North American Free Trade Agreement (NAFTA), 62, 126, 130

North Carolina Metropolitan Coalition, 85
North Sea (conference regarding), 78
Norton, Gale, 89, 146-147
"not in my backyard," 8, 104
novelty, newsworthiness criterion, 57

O

ocean water quality, 57
Office of Management and Budget, 25, 35, 43, 114
Office of Technology Assessment (OTA), 69
office terms, 20-21
oil spills, 16, 44, 54, 112, 115, 116
old-growth redwood forests, 57
Olin, 119
online activism, 137-151
Open Secrets, 143
Organic Consumers Association, 107
organized labor, 11, 123
Osborne, David, 48
Our Stolen Future, 18
overpopulation, 106
overregulation, 114
ozone layer depletion, 58, 71, 79

P

Pacific Corporation, 26
Pacific Gas and Electric, 123
Pacific Institute for Public Policy, 120
Pacific Legal Foundation, 120
packaging eco-labeling, 132
PACs. *See* Political action committees (PACs)
partisanship, 2-4
PCBs. *See* Polychlorinated biphenyls (PCBs)
peace, relationship to environmental problems, 136
peacekeeping missions, Haiti, 22
peer review, 66-67, 69, 71, 74
People of Color Leadership Summit (First National), 49, 100
Persistent organic pollutants (POPs), 143, 147
pesticides, 6, 71
petitions, 149
photographic depiction books, 93-94
Physicians for Social Responsibility, 75
Pinchot, Gifford, 5, 118
"planetary health," 9
"The Poisoning of America," 55

Polaroid, 116
policymaking by decree, 25-29
policy neutrality, 36, 76
Political action committees (PACs), 19, 142
Political Economy Research Center, 119
pollution, 15, 43, 71, 86, 113. *See also*
    specific pollutants
pollution credits, 96, 123
The Pollution Prevention Act of 1990, 79
polychlorinated biphenyls (PCBs), 98, 147
POPs. *See* Persistent organic pollutants
    (POPs)
population interests, 36, 58, 84, 103
pork barrel projects, 22
post-World War II era, 14-15, 93, 113
Precautionary principle, 78-80, 133, 158
presidential office
    legislation process, 17
    rule by "decree," 25-26
    website, 138-139
President's Council on Sustainable
    Development, 79
Price, Tom, 137
"Principles of Environmental Justice," 49
privatization, 45, 115
Project on the State of the American
    Newspaper, 82
Project XL (Excellence and Leadership),
    49-50
prominence, newsworthiness criterion, 57
proximity, newsworthiness criterion, 57
public health threats, 6-7, 15, 61
"public injuries," 14
public review process, 36
public views
    business community activities, 111
    Congress being "government," 82
    environmental science, 77
    EPA, 48
    general politics, 1-2
    global climate change, 131
    media coverage, 55-56, 61
    unaware of think tank sponsors, 120

Q

"Quality of Life Reviews," 43
Quayle, Dan, 45
quotations
    anonymous British delegate, 65
    Bailey, Doug, 137

Bierce, Ambrose, 1
von Bismarck, Otto, 13
Castells, Manuel, 81
The Competitive Enterprise Institute, 111
Dingell, John, 31
Gibbs, Lois, 91
Gifford, Bill, 91
Hammonds, Tim, 65
Latham, Earl, 31
Price, Tom, 137
Rogers, William H., 13
Schneider, Stephen H., 53
Schultz, Ernie, 53
Spitzer, Eliot L., 81
Turning Point Project, 129
Wasow, Bernard, 129
Woolard, E.S., 111

R

racism. *See* Minority concerns
radioactive waste, 98
radon infiltration, 7, 15, 61, 77
Rainforest Action Network, 107, 146
rain forests, 8
Ray, Dixie Lee, 63, 74, 75
Reagan administration, 33, 40, 47, 83, 95,
    113
Reason Foundation, 74, 119
Redford, Robert, 57
regneg (negotiated rule-making), 41
regulations
    adoption, 35-36
    legislation comparison, 32-35, 34
    policy implications, 154
    process, 32-33, 40
    ultra vires, 39
Regulatory Analysis Review Group, 43
Regulatory Improvement Act of 1999, 50
Regulatory Right-to-Know Act of 1999, 50
Reilly, William, 100
Reinventing Environmental Regulation,
    41-42, 50, 154
Reinventing Government, 48
religious groups, 9
representation, organizations, 9
Republicans, 1, 24, 69, 73
responsibility for safe management, 7
Responsible Care code of conduct, 115-116
riders, 21-26, 40, 143
"right to know," 126, 139

Right wing, 1
Rio Conference on the Environment and
        Development Declaration, 133
risk assessment, 44, 50, 69, 70
Rogers, William H., 13
Rohrabacher, Dana, 70
Roosevelt, Theodore, 26
Ruckelshaus, William, 43, 45

S

Sagebrush Rebellion, 88, 114, 118
Save Our Environmental Action Center, 146
SayNoToNorton.com, 147
Scaife, 119
Schneider, Stephen H., 53
Schultz, Ernie, 53
Science
    background, 156
    community, 10
    Congress, relationship with, 68-72
    courts, relationship with, 72-73, 158
    fringe science, 70-71
    Internet effect, 148
    "junk science," 73, 75-76
    lawmakers, relationship with, 66-68
    media, relationship with, 73-76, 156
    regulatory level, 72
    science policy oxymoron, 73
    scientific truth, 71
    "sound science," 66, 69-70, 72, 121
    ventriloquism through media, 76
Science, 74, 123
Science and Environmental Health
        Network, 79
Science and Environmental Policy Project,
        74, 119
Science Under Siege, 66, 75
Scientific American, 55
"Scientific Integrity and Public Trust: The
        Science Behind Federal
        Policies and Mandates," 70
Scorecard, 140
sea turtles, 133
Sector Facilities Indexing Project, 139
see-sawing, scientific opinion, 77
self-interest, 5, 39
Senate, 16-17, 19, 47
sequoia groves, 28
Setting Priorities, Getting Results, 46
"sick building syndrome," 7

Sierra Club, 26, 91, 93-94, 95, 146
Sierra Legal Defense Club, 92
Silent Spring, 6-7, 15, 18, 54, 109
silicon breast implants, 76
Silicon Valley Toxics Coalition, 107
Simon, Julian, 63
Singer, Fred S., 74
Skull Valley Band of the Goshute, 4
Small Business Association, 112
The Small Business Regulatory Enforcement
        Fairness Act, 40
Smith, Benjamin, 146
smog, 123
social justice, 97-105
solid waste reduction and disposal, 15
"sound science," 66, 69-70, 72, 121
sources, for media, 59-60
spending bills, 22
Spitzer, Eliot L., 81
standards
    emission, 87
    establishing minimum national, 82
    fuel economy, 24-25, 123
    gasoline, 133
    lowest common denominator, 133
    states satisfying, 85
standing reference committees, 16-17
state governments
    collaboration to enforce standards, 87
    federal laws and policies, 37, 82-86, 105
    influence, 11
    as lobbyists, 81-90, 158-159
    pollution crossing state boundaries, 86,
        87
    regulatory burden, 37
    staffing decline, 82
"The State of the Earth," 55
steering committee, 35
Stossel, John, 63
strategy services, 18
Streep, Meryl, 57
Styrofoam, 77, 97, 123
Subcommittee on Energy and Environment
        of the House of Representatives
        Committee on Science, 70
suburban sprawl, 8
sugar production, 88
summaries of Central Ideas
    business-driven environmental politics,
        127
    changes in environmental politics, 12

environmental advocacy, 110
environmental regulation, 51
international environmentalism, 136
legislation, 29
media, 64
online activism, 151
science, 80
state and local governments, 90
Sun Company, 116
Superfund sites, 15, 32, 33, 61
surrogate lobbying, 117-120
"sweetheart" deals, 88

T

tabling bills, 19
Task Force on Regulatory Relief, 44
Taylor, Charles, 23
telecommunication technology. *See*
    Internet
"10 Principles for Reinventing
    Environmental Regulation," 50
terms of office, 20-21
thecampaign.org, 146
"Think globally, act locally," 130
think tanks, 74-75, 118-120
Thomas, William, 29
threatened species, 15
Three Mile Island, 16, 33, 44, 54, 98, 112,
    115
timeliness, newsworthiness criterion, 57
Time magazine, 55
Times Beach, 16, 44, 54, 98
Tongass National Forest, 3, 23
Toxic Catastrophe Prevention Act, 3
Toxic Release Inventory, 139
toxic substances, 15, 122
Toxic Waste and Race in the United States,
    99-100
trading votes, 19
Train, Russell, 43
Trashing the Planet, 75
The True State of the Planet, 75, 120
tuna, 133
Turning Point Project, 106-107, 129
turtles, 133

U

*Unfinished Business—A Comparative
    Assessment of Environmental
    Priorities*, 46
Unilever, 123
Union of Concerned Scientists, 75
United Auto Workers, 126
United Mine Workers, 4
United Nations Conference on
    Environment and Development,
    78
United Nations Framework Convention on
    Climate Change, 79
United States Conference of Mayors, 85
United States Supreme Court, 89
*Unlocking Our Future: Toward a New
    National Science Policy*, 71
unregulated market promotion, 74
urban redevelopment, 36
U.S. Bureau of Mines, 69
U.S. Chamber of Commerce, 112, 116
U.S. Fish and Wildlife Service, 19, 22, 88
U.S. Forest Service, 3, 88
U.S. Geological Survey, 69
Utah Association of Local Governments, 27
Utah Wilderness Alliance, 26

V

Valdez Principles, 116
vehicles, scrapping, 123
ventriloquism of scientists, 76
Veterans Affairs—Housing and Urban
    Development appropriations
    bill, 24
veto authority, 17, 46
victim stance of environmental concerns,
    97-104
violence, 107-109
votes, trading, 19

W

Wal-Mart, attack on, 106
Washington Legal Foundation, 75
Wasow, Bernard, 129
water, drinking, 7, 15
Watson, Robert T., 69
Watt, James, 88, 89, 95, 96, 118
Waxman, Henry, 24

websites. *See also* Internet
    Center for Responsible Politics, 142-143
    Dirty Money Tracker, 143
    Envirofacts Warehouse, 139
    Environmental Defense, 140
    Environmental Monitoring for Public
        Access and Community
        Tracking, 139
    Environmental Protection Agency (EPA),
        139
    Environmental Working Group, 140-141
    Fox News, 76
    League of Conservation Voters, 154
    members of Congress, 138
    NoMoreScares.com, 75
    Open Secrets, 143
    SayNoToNorton.com, 147
    Scorecard, 140
    Sector Facilities Indexing Project, 139
    Surf Your Watershed, 139
    thecampaign.org, 146
    Toxic Release Inventory, 139
Western States Coalition, 27
wetlands, 15, 33-34, 36
Whitehouse, 138-139, 149
Wildavsky, Aaron, 63
Wilderness Society, 91, 95, 146

Wilkinson, Todd, 66
Wise Use, 118-120, 160
Woburn cancer cluster, 98
Woodcock, Leonard, 126
wood products, 123
Woolard, E.S., 111
"worker right to know," 126
World Bank, 135
world peace, relationship to environmental
        problems, 136
World Trade Organization (WTO), 4,
        132-135
post-World War II era, 14-15, 93, 113
Worldwatch Institute, 75
World Wildlife Fund, 123, 146
WTO. *See* World Trade Organization
        (WTO)
Wyoming, 4

Y

Yosemite National Park, 6

Z

Zero Population Growth (ZPG), 106, 107,
        146